Brain Death

Brain Death
A Reappraisal

Calixto Machado

Department of Clinical Neurophysiology
Institute of Neurology
Havana, Cuba

 Springer

Calixto Machado
Department of Clinical Neurophysiology
Institute of Neurology and Neurosurgery
Apartado Postal 4268
Havana, Cuba 10400
braind@infomed.sld.cu

Library of Congress Control Number:

ISBN-13: 978-0-387-38975-2   e-ISBN-13: 978-0-387-38977-6

Printed on acid-free paper.

9 8 7 6 5 4 3 2 1

springer.com

To my dad, who instilled in me since childhood, a passion for medicine and science.

To my mom, who made home a place filled with love and happiness in which my sister and I learned to grow up as good human beings.

To my wife, physician and poet, my life mate since I was a child, who invented for us a magic kingdom from which I would try to conquer the mysterious territories of neuroscience and those even more profound of my own existence.

To my two amazing daughters, my most valuable treasures, who also decided to walk the thorny road to become medical doctors.

To my dog, Pituzo, for being an everlasting source of happiness.

Preface

Historically, long before modern technology, death was considered to have occurred when heartbeat and breathing ceased, and the soul abandoned the body. Nonetheless, the concept of death evolved as technology progressed, forcing medicine and society to redefine the ancient cardiorespiratory diagnosis to a neurocentric diagnosis of death.

In the meantime, the technical and scientific advances of the last century provided effective mechanical ventilators and cardiopulmonary resuscitation ("reanimation"), which compelled physicians in the late 1950s to confront a condition impossible even to imagine previously—a state in which the brain is massively damaged and nonfunctional while other organs remain functioning. Was such a patient alive or dead? This radically changed the course of the debates about human death, marking a turning point when brain-oriented definitions of death started to be formulated, and brain death (BD) was gradually accepted as death of the individual.

However, it is commonly believed that the concept of BD evolved to benefit organ transplantation. A historical approach demonstrates that both BD and transplantation had fully separate origins. Organ transplantation became possible with technical advances in surgery and immunosuppressive treatment. The concept of BD evolved with the introduction of intensive care units.

Moreover, Terry Schiavo and other famous patients, including Karen Ann Quinlan and Nancy Cruzan in the United States and Tony Bland in the United Kingdom, raised new controversies about the diagnosis and management of the persistent vegetative state (PVS) and the minimally conscious state (MCS), with powerful ethical and legal debates about diagnosing these patients as dead, or stopping fluids and nutrition, or keeping the patients alive, attempting to rehabilitate them from the terrible clinical conditions.

In this book I describe my experience of more than 20 years devoted to brain death and altered consciousness, with a detailed literature review. If this book promotes discussion and serves as a source of information in this area, I will be satisfied.

I have studied death because of my love of life and human beings. All humans should have the right to live and die with dignity!

Calixto Machado, M.D., Ph.D
Havana, Cuba
December 13, 2007

Contents

1
The Concept of Brain Death Did Not Evolve to Benefit Organ Transplants

Before the advent of modern technology, death was considered to have occurred when the heartbeat and breathing ceased, and the soul abandoned the body. The absence of fog on a glass or a mirror placed under the nostrils and the patient's failure to get up after being called three times by name were popular methods to document death.[1,2]

In the meantime, the noteworthy technical and scientific advances of the last century provided the possibility to develop effective mechanical ventilators and cardiopulmonary resuscitation. At the end of the 1950s some reports appeared about a condition impossible even to imagine previously: one in which the brain is massively damaged and nonfunctional while other organs remain functioning. Was such a patient alive or dead? This radically changed the course of the debates about human death, marking a turning point when brain-oriented definitions of death started to be formulated, and brain death (BD) was gradually accepted as death of the individual.[2,3]

However, it is commonly believed that the concept of BD evolved to benefit organ transplantation.[4–6] A historical approach demonstrates that BD and organ transplantation had fully separate origins. Organ transplantation became possible with technical advances in surgery and immunosuppressive treatment. The concept of BD evolved with the development of intensive care.[7]

This chapter discusses the development of the concept of BD in the 20th century for comparison with transplant advances. We will see that although nowadays both BD and organ transplants are fully integrated, the concept of BD did not originate to benefit transplantation.

Historical Evolution of the Brain Death Concept

During the 18th and 19th centuries, resuscitation and artificial respiration raised doubts about the differences between life and death. Physicians and researchers had identified disorders that mimicked death including suspended animation induced by hypothermia, intoxication by opiates or alcohol, suffocation, apoplexy (stroke), high fever, head injury, and anesthesia. Even more controversial were states of

FIGURE 1.1. René Théophile Hyacinthe Laënnec (1781–1826) invented the stethoscope in 1819. (Courtesy of the National Library of Medicine.)

psychological or idiopathic suspended animation, such as catalepsy, and hypnotic trance. Richardson[8,9] listed drugs that could induce apparent death, including atropine, amylnitrate, atropine, curare, chloral hydrate, cyanide, and pure oxygen. For this author catalepsy was the result of endogenously produced amylnitrate.[2,8,9]

The French researchers also supported a cardiorespiratory view of life and death. Rene Laënnec (Fig. 1.1) invented the stethoscope in 1819,[10–17] and another French physician, Eugene Bouchut,[18] first applied the new device to the diagnosis of death in 1846. With these contributions the public's fear of premature burial decreased to some extent.[2]

Resuscitation and Techniques for Restoring or Substituting Heartbeat and Lung Functions

The 20th century's advances in resuscitation made a concept of death seem even more indefinable. After about 1900, apnea and absence of heartbeat in the operating

FIGURE 1.2. Vladimir A. Negovskii (1909–2003). Father of reanimatology. (Courtesy of the National Library of Medicine.)

room were often effectively reversed by intermittent positive-pressure ventilation by bellows or bag or open-chest cardiac massage. Also, three generations of Russian and Soviet scientists, from Bachmetieff in 1910 to Negovskii (Fig. 1.2) up to the 1960s, provided astonishing descriptions of reviving the apparently dead from suspended animation and resuscitation.[19–22] Similar contributions came from the United States and France, led by Peter Safar (Fig. 1.3), the father of the modern resuscitation.[23–26] The discovery that deep hypothermia made it possible for animals and even humans to withstand an hour or more without pulse or respiration, and the introduction of effective cardiopulmonary resuscitation, progressively led to the conclusion that the main target organ to protect after cardiorespiratory arrest is the brain.[27–32]

Another invention to recover heartbeat after asystole was the cardiac defibrillator. At the end of the 1920s, European and American physicians, learning from accidental electrocution, applied electric stimulation to reverse ventricular fibrillation.[2] Nonetheless, it was only in 1947 when Claude Beck, a pioneering cardiovascular surgeon in Cleveland, directly defibrillated a human heart during cardiac surgery with success.[33] In 1956 Paul Zoll, a cardiologist, used a more powerful defibrillator and carried out closed-chest defibrillation.[34] Cardiac defibrillators progressively became keystone instruments in emergency services and intensive care units.

Another invention that undermined the physician's ability to diagnose death was mechanical respiration, starting approximately 50 years ago in an epidemic of poliomyelitis. Drinker described external negative pressure to the chest wall. This

FIGURE 1.3. Peter Safar during the III International Symposium on Coma and Death, Havana, 2000. Father of modern reanimatology. (See also color insert.)

"iron lung" dramatically reduced mortality from poliomyelitis. In the early 1930s, John Emerson, a Massachusetts inventor, improved the apparatus, and iron lungs were distributed by the thousands throughout the Western world.[35–40]

With the eradication of polio in the 1950s and 1960s, and the invention of portable ventilators, the iron lung became obsolete. Partly because of the invasive nature of the tracheostomy often needed for tank ventilators, positive pressure ventilators were introduced and became the standard.[35–43]

Gibbon first used a mechanical extracorporeal oxygenator, or cardiopulmonary bypass (CPB), which he termed the heart-lung machine. On May 6, 1953, Gibbon performed the first successful open heart operation using a heart-lung machine while repairing an atrial septal defect. In doing so, the heart-lung machine was used in the venoarterial mode to provide total support or to substitute for heart and lung, to facilitate cardiac operations.[44,45] Settergren has described the introduction of CPB in Sweden in 1954, one year after Gibbon's first operation, when the heart was stopped in the operating room; the moment of death was noted as the time when the CPB was turned off. When CPB was still functioning, patients had small and reactive pupils; when CPB was disconnected, the pupils became mydriatic and nonreactive. As Settergren commented, this was an unintentional introduction to the concept of BD in cardiac surgery in Sweden.[46]

Historical Accounts of Brain Death

The earliest accounts concerning states resembling what would today be recognized as BD go back to the end of the 19th century, when several authors reported that, during an increase of intracranial pressure (ICP) in experimental models and in patients, respiration suddenly stopped, whereas the heart continued to beat.[47-54]

In 1892, Jalland[49] reported a case with a brain abscess, stating: "The marvelous way in which respiration returned after having ceased for some time, as the pus was evacuated, was the cause of surprise to those who were present at the operation." Horsley[51] stated in 1894 that patients suffering from cerebral hemorrhage, brain tumors, and depressed fractures of the skull "die from respiratory and not from cardiac failure." Duckworth.[52] reported "some cases of cerebral disease in which the function of respiration entirely ceases for some hours before that of the circulation." Finally, Harvey Cushing [53,54] (Fig. 1.4) stated that, "in death from a fatal increase in intracranial tension the arrest of respiration precedes that of the heart .... Prompt surgical relief, with a wide opening of the calvarium, may save life even in desperate cases with pronounced medullary involvement." Although these reports provide early descriptions of the BD syndrome,[47-54] their authors still considered cessation of respiration or heartbeat to be the sign of death.

In 1929, Berger first reported and recorded the electroencephalogram (EEG) in a human being, which he termed, in German, *Elektroenkephalogramm*[55-63]; he

FIGURE 1.4. Cushing and a young a patient in 1928. (Courtesy of the National Library of Medicine.)

FIGURE 1.5. George Washington Crile. (Courtesy of the National Library of Medicine.)

said that July 6, 1924, was the date of discovery. Neurophysiologists had gained the ability to measure a function of the brain directly, without relying on heart or lungs.

Within a year of Berger's report, George Washington Crile (Fig. 1.5), a urologist, remarked that "processes leading to death were always accompanied by a decrease in the conductivity of the brain and of other components of the central nervous system." He stated that the cause of clinical death was the decrement of potentials because the loss of membrane potentials caused the death of single cells or of tissue cultures. Hence, he provided a concept of life and death: "Life may be defined as a potential which is maintained and is varied adaptively according to environmental conditions, this potential being maintained by chemical activity— mainly by oxidation. The loss of this potential is death." He also noted that life in unicellular organisms depends on the electric potential between the cytoplasm and the surrounding medium and between the cytoplasm and the nucleus; in lower multicellular organisms life is derived from the electric potential difference between the central nervous system (CNS) and the rest of the organism, while in the highest forms of multicellular organisms life is based on an adaptive difference in electric potential between the brain and other organs and tissues, especially the liver.[64,65]

In 1938, Sugar and Gerard[66] temporarily occluded the carotid artery in cats and demonstrated with the EEG that abolition of electric potentials in different brain regions correlated with their resistance to ischemia. In 1939, Crafoord[67] stated that death was due to "cessation of blood flow to the brain and nothing else." In 1900, Hill[68] concluded from experiments that "total deprivation of blood immediately paralyzes the brain." These were striking accounts concerning the importance of cerebral blood flow in maintaining brain function.

At the end of the 1950s neuroradiologists and neurosurgeons recognized *cerebral* circulatory arrest. This angiographic picture was identified in apneic or comatose patients after head injury, intracranial hemorrhage, or other space occupying brain lesions.[46] Riishede and Ethelberg[69] in 1953 had first reported this radiological finding in coma, but they attributed the pattern to brainstem reflexes changing the vascular tone. Löfstedt and von Reis,[70] in 1956, described six patients without using contrast material in the cerebral circulation. The patients were being treated with a ventilator after respiratory arrest; they were unconscious with no central reflexes, hypotensive, and would be hypothermic if not treated with vasopressors. The patients "died" (had circulatory arrest) after a few days. At autopsy there was no obstruction of the cerebral arteries. The authors wrote that cerebral circulation might have been too slow to be shown by angiography (Fig. 1.6), but they concluded that increased intracranial pressure and vasospasm were the most

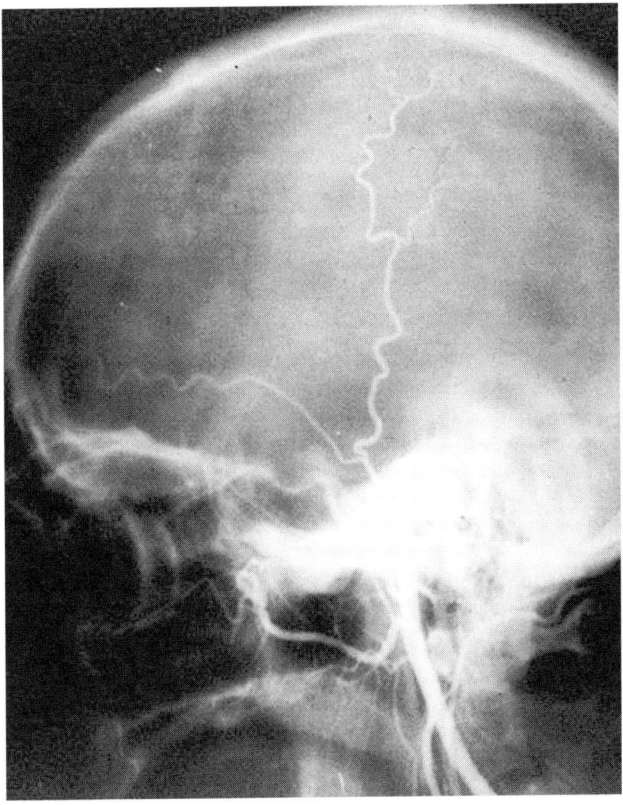

FIGURE 1.6. Cerebral angiography showing arrest of intracranial circulation. (Courtesy of Dr. Esperanza Barroso.)

FIGURE 1.7. Pierre Wertheimer. (Courtesy of the National Library of Medicine.)

probable explanation for the angiographic findings. They described similar patients in 1959.[46,71]

Two landmark accounts appeared in 1959. At the beginning of that year, Wertheimer's group[72,73] characterized the "death of the nervous system" (Fig. 1.7). Some months later, Mollaret and Goulon[74] coined the term *coma dépassé* for an irreversible state of coma and apnea.

Wertheimer's group[72,73] described "the death of the nervous system" and equated the condition with the heart-lung preparation of the physiologists. The group went further to propose stopping the ventilator if death of the nervous system was diagnosed clinically and by "the repeatedly verified absence of electroencephalographic (EEG) activity both in the cortex and in the diencephalon, and if resuscitative efforts have been given enough time, 18 to 24 hours" (translation by Settergren[46]).

Based on 23 cases, Mollaret and Goulon described a condition of deep coma with no spontaneous respiration, no reflexes, polyuria, hypotension if norepinephrine was not given continuously, and the absence of all EEG activity. They stated that if the ventilation or the infusion of norepinephrine were stopped, the patient would rapidly die, and that cardiac arrest would follow. Despite treatment the patients died from cardiac arrest usually within days after the deep coma had been established.[74] Although Mollaret and Goulon's article was a breakthrough contribution to the description of brain death, they did not believe in the death of the nervous system as understood by Wertheimer et al.[72] They considered the concept of cardiac death to be the defining phenomenon, generating a major controversy.[72,74]

TABLE 1.1. Criteria for establishing brain death (proposed by Schwab et al.[76] in 1963)

1. Spontaneous respiration must be absent for at least 30 minutes.
2. There must be no tendon reflexes of any type.
3. There must be no pupillary reflexes and the pupils must be dilated.
4. Eyeball pressure must not change the heart rate.
5. The EEG should show flat lines without any rhythms in all leads over a 30-minute period.
6. A loud noise must not produce any detectible discharge in the EEG.
7. The inter-electrode resistance is usually over 50,000 Ω.

Guy Alexandre, a Belgian surgeon, in 1963 adopted BD diagnostic criteria similar to those from the Harvard Committee and applied those criteria in performing the first organ transplant from a brain-dead donor (see Chapter 2).[75]

In the same year, Robert S. Schwab et al.[76] proposed that "the total absence of EEG activity after 30 min of recording is the most important evidence of death of the nervous system." They were also among the first to accept the diagnosis of human death on neurological grounds in the presence of a beating heart. They referred to these cases as "human heart-lung preparation." They went further than just recommending EEG as an ancillary test: they incorporated EEG findings into their diagnostic criteria, which were similar to those of the Harvard Committee (Table 1.1).[76]

Schwab was later an expert of the Harvard Committee. Nonetheless, it seems that the Harvard Committee participants were not aware of the French contribution in this area. Hence, there were parallel approaches to the subject of BD.[77]

Settergren[46] also chronicled that in 1965, Swedish neurosurgeon Frykholm circulated a memorandum at a meeting arranged by the Swedish Medical Authorities on the rules for transplantation, suggesting that that the ventilator should be stopped in patients with no reflexes from the cranial nerves, an isoelectric EEG, and with no blood flow through the brain at angiography. He tried to implement it in organ transplantation, arguing that this definition of death, if it were to be accepted by the public, had to be included in a law and that if these dead patients were suitable as donors for organ transplants, this could justify continuing ventilation. As had happened to Wertheimer, Frykholm was criticized for this opinion.[46]

Two major medical events occurred in 1968, a crucial year for the development of the BD concept. On August 5, the 22nd World Medical Assembly met in Sydney (Australia) and released an essential statement on human death. And due to the caprices of history, the *Journal of the American Medical Association (JAMA)* published a landmark article: "A Definition of Irreversible Coma: the Report of the Ad Hoc Committee of the Harvard Medical School to Examine the Definition of Brain Death."[78]

The Declaration of Sydney touched philosophical issues about death, reporting that "death is a gradual process at the cellular level with tissues varying in their ability to withstand deprivation of oxygen," and stating that death "lies not in the preservation of isolated cells but in the fate of a person."[79]

The Harvard Report had momentous repercussions and constituted a break-through account, establishing a paradigm for defining death by neurological criteria. After the Harvard Report, BD was widely accepted.[3]

History of Transplantation

Ulmann[80] (Fig. 1.8) in January 1902, reported the first renal autotransplantation in a dog: grafting a kidney into the neck. In 1906, Jaboulay,[81] connected sheep and pig kidneys, respectively, to the brachial vessels of two patients who were dying of renal failure. Neither kidney worked, but these were the first transplants, albeit xenografts, that had been placed in humans. The techniques used to join the vessels together were those developed and described by Carrel[82,83] (Fig. 1.9), who is known as the founding father of experimental organ transplantation because of his pioneering work with vascular techniques.

In 1936, the Soviet surgeon Voronoy[84] reported the transplantation of a human kidney to a patient suffering from acute renal failure. There was a major mismatch of blood groups, and the kidney never functioned. Voronoy carried out six kidney transplants without success.[84–86]

FIGURE 1.8. Emerich Ulmann reported the first renal autotransplantation in a dog. (Courtesy of the National Library of Medicine.)

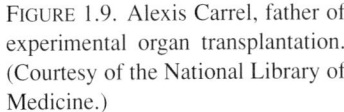

FIGURE 1.9. Alexis Carrel, father of experimental organ transplantation. (Courtesy of the National Library of Medicine.)

The modern era of clinical transplantation began in the early 1950s, highlighted by the contributions of Kuss, Dubost, Servelle, and Michon and their associates in France,[87–90] and by Hume et al.[91] in Boston. Hume made an early attempt at allografting the kidney from an unrelated donor and the kidney functioned well, but only for a short period.

In 1954, Murray (Fig. 1.10A) and his team[92,93] carried out the first successful human organ transplant, taking a kidney from an identical twin (Fig. 1.10B). This was a landmark event in the history of transplantation.

In 1962, Murray performed the first successful cadaveric kidney transplant.[94] In 1963, Starzl[95] achieved the first human liver transplant, and Hardy et al.[96] performed the first lung transplant. In 1966, Lillehei and Kelly's group[97,98] carried out the first successful pancreas transplant. The development of immunosuppressive treatment played a crucial role in those early successful organ transplants attempts between unrelated recipients and donors.[99]

Another milestone event in the history of transplantation occurred in South Africa, on December 3, 1967, when Barnard (Fig. 1.11) carried out the world's first successful human heart transplant.[100–108]

In those days, the surgical team brought a brain-dead donor into the operating room with the recipient for the removal; the respirator was then stopped, and

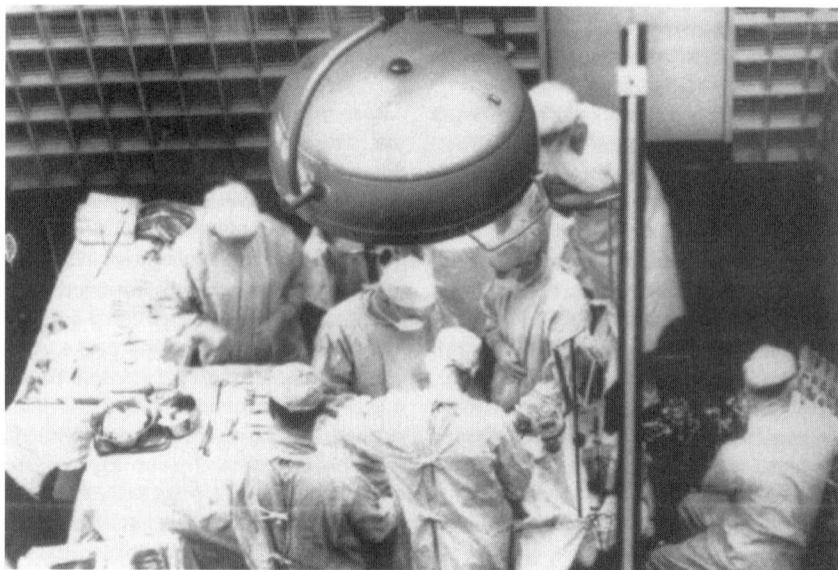

FIGURE 1.10. (A) Joseph E. Murray honored at a special commemorative event celebrating 50 years of transplantation at the U.S. Transplant Games on July 29, 1994. (B) Murray and his team carried out the first successful human organ transplant, taking a kidney from an identical twin on December 23, 1954. Photos credit Eric Miller for the National Kidney Foundation, and the Surgery of the Soul, Science History Publications, Watson Publishing International. (See also color insert.)

FIGURE 1.11. Christiaan Barnard. Photograph taken during the proceedings of a meeting held on December 28, 1967, at O'Hare Airport, Chicago, Illinois, on the subject of cardiac transplantation. Left to right: David M. Hume, Christiaan Barnard, Andrew G. Morrow, Donald S. Fredrickson. Courtesy of the National Library of Medicine. (See also color insert.)

everyone waited for the donor's heart to cease to beat. Technically, therefore, those donors were *not* brain dead at the time of organ retrieval. Rather, they had been declared dead by classic cardiorespiratory criteria.[75]

Even when Barnard reported his first successful heart transplant, he stated: "As soon as it had become obvious that, despite therapy, death was imminent in the donor, the recipient was anaethesized." He continued: "As soon as the patient had been certified dead (when the electrocardiogram had not shown activity for 5 minutes and there was absence of any spontaneous respiratory movements and absent reflexes)."[100] When Barnard was celebrating the 20th anniversary of his first heart transplant, he stated:

"Here I consulted the Professor of Forensic Medicine at University of Cape Town (UCT), Professor L.S. Smith, who informed me that the legal authorities had wisely left the definition of death imprecise, and that, if a doctor so deemed, death of the brain was legally acceptable as evidence of death. (Nevertheless, to be quite sure that I did not run into legal problems, in the first heart transplant operation I disconnected the respiratory support being given to the potential donor and waited until the heart stopped beating before removing it and transplanting it into the recipient.)"[103]

Dr. Raymond Hoffenberg participated in Barnard's second heart transplant. He commented: "On my last night as the consultant on-call I was asked by the transplant team to pronounce the man 'dead' and confirm that his heart would be suitable for transplantation." He continued:

My patient still had a few elicitable neurological reflexes. I went home, returned an hour or two later, still found the reflexes, and declined to pronounce him dead. . . . I said I could not agree to the removal of the heart from someone who still showed signs of 'life,' and then spent a sleepless night wondering whether I was being unnecessarily obstructive.... I went to the hospital very early next morning and satisfied myself that I could no longer elicit the reflexes, and the surgery went ahead.

It is obvious that although Dr. Hoffenberg had not applied a defined set of BD diagnostic criteria, he was nevertheless referring to "signs of life," the preservation of some neurological reflexes.[105]

Summary and Conclusions

Table 1.2 lists the chronology for transplants and the brain death concept in the 20th century. At the beginning of that century, when the first renal autotransplant was carried out, there were also several reports of increased ICP in experimental models and patients with ensuing respiratory arrest but with preserved heartbeat.[47–54] From 1902 to 1950, development of the BD concept was supported by the discovery of EEG,[47–54] Crile's definition of death,[65] the use of EEG to demonstrate abolition

TABLE 1.2. Keynote historical moments of the development of the brain death concept and transplants during the 20th century, in different periods, are comparatively presented (see text for explanation)

Transplants	Brain death concept
1902: First renal autotransplant	1902: Reports of increased ICP provoking respiratory arrest with preserved heartbeat
1902–1950	1902–1950
1906: First xeno-transplant in humans	1929: Berger discovers the EEG
1933: First kidney transplant from cadaveric donor, without success	1930: Crile proposes a definition of death as a drop in electric potentials
	1938: EEG is used to show loss of brain potentials after ischemia
	1939: Crafoord proposes that death in animals is due to cessation of cerebral blood flow
1950–1959	1950–1959
1954: First successful kidney transplant between identical twins	1956: Cerebral circulatory arrest is demonstrated in comatose patients by angiography
	1959: Death of the Nervous System Coma dépassé
1960–1968	1960–1968
1962: First successful kidney transplant from a cadaveric donor	1963: First organ transplant using a brain-dead donor
1963: First successful liver and lung transplant from a cadaveric donor	Schwab proposes EEG to demonstrate death of the CNS
1966: First successful pancreas transplant from a cadaveric donor	
1968: First successful heart transplant between humans	1968: Harvard Report and Sydney Declaration

of brain electrical potentials after induced ischemia,[66] and Crafoord's statement that death was due to "cessation of blood flow."[67] This epoch was underscored in transplantation by the first xenotransplant in humans (1906),[80] and the first kidney transplant from a cadaveric donor (1933).[84] From 1950 to 1959, several articles appeared about cerebral circulatory arrest detected by angiographic studies of comatose patients.[46,69–71,109,110] Moreover, two crucial accounts emerged in 1959: descriptions of the death of the nervous system and coma *dépassé*.[72–74] During the same period, Murray performed the first successful kidney transplant.[92] In 1963, an instantaneous conjunction took place between the evolution of the BD notion and transplantation, when the first kidney transplant using a brain-dead donor was carried out.[75] In the same year, Schwab proposed the use of EEG in establishing BD.[76] In 1968, two landmark arguments for accepting BD appeared: the Harvard Committee Report[78] and the Declaration of Sydney.[79] During this period, the first successful kidney, lung, and pancreas transplants using cadaveric (not brain-dead) donors were performed.[94–97] Transplantation history was highlighted in 1967 with the first heart transplant between humans.[100]

This historical review demonstrates that the BD concept and organ transplantation had different origins and followed a parallel progression, which had an instantaneous conjunction in 1963, and finally joined and began to move along together after the Harvard Committee Report.[78] Hence, the BD concept did not evolve to benefit transplantation.

I speculate that if, in the future, the need for organs for transplantation from brain-dead donors is fully substituted by xenotransplants, the manufacture of tissues and organs by stem cell–derived techniques, or by other methods not yet known, the BD concept and transplantation will resume a parallel development.

In any event, in that time to come, when a patient is receiving life support, the physician will still need to face the thorny clinical state created by a nonfunctioning brain and a preserved heartbeat.

References

1. Pallis C. Brainstem death. In: Braakman R, ed. *Handbook of Clinical Neurology: Head Injury*. Amsterdam: Elsevier Science B.V., 1990:441–496.
2. Pernick M. Back from the grave: recurring controversies over defining and diagnosing death in history. In: Zaner RM, ed. *Death: Beyond Whole Brain Criteria*. Dordrecht, Netherlands: Kluwer Academic, 1988:17–74.
3. Wijdicks EF. Brain death worldwide: accepted fact but no global consensus in diagnostic criteria. *Neurology*. 2002;58:20–25.
4. Siminoff LA, Burant C, Youngner SJ. Death and organ procurement: public beliefs and attitudes. *Kennedy Inst Ethics J*. 2004;14:217–234.
5. Youngner SJ. Brain death and organ transplantation: confusion and its consequences. *Minerva Anestesiol*. 1994;60:611–613.
6. Youngner SJ. Defining death. A superficial and fragile consensus. *Arch Neurol*. 1992;49:570–572.
7. Machado C. Consciousness as a definition of death: its appeal and complexity. *Clin Electroencephalogr*. 1999;30:156–164.

8. Richardson BW. Suspended animation. *Living Age.* 1879;142:99–103.
9. Richardson BW. The absolute signs and proofs of death. *Proc Med Soc Lond.* 1888;12:100–117.
10. [Laënnec discovered the stethoscope while seeing children play; his discovery made the diagnosis of heart & chest diseases possible.] *Clin Lab (Zaragoza).* 1957;64(suppl).
11. Adler C. Rene Theophile Hyacinthe Laennec (1781–1826), inventor of the stethoscope. *Adler Mus Bull.* 1981;7:3–7.
12. Bloch H. The inventor of the stethoscope: Rene Laënnec. *J Fam Pract.* 1993;37:191.
13. Kerridge IH. Rene Laënnec and the introduction of the stethoscope. *Aust N Z J Med.* 1996;26:407–410.
14. Logan JS. An autograph letter of Dr. Rene Laënnec. *Ulster Med J.* 1972;41:108–110.
15. Sergent E. [Laënnec as physician.] *Dia Med.* 1950;22:3430–3433.
16. Somolinos-Palencia J. [Rene Theophile Hyacinthe Laënnec, on the 200th anniversary of his birth.] *Gac Med Mex.* 1981;117:122–126.
17. Tan SY, Yeow ME. Rene Laënnec (1781–1826): inventor of the stethoscope. *Singapore Med J.* 2005;46:106–107.
18. Bouchut E. *Traité des Signes de la Mort et des Moyens de Prévenir les Enterrements Prématurés.* Paris: Bailliere, 1849.
19. Negovskii VA, Gurvich AM. [The development of the science of resuscitation in the USSR.] *Patol Fiziol Eksp Ter.* 1967;11:41–48.
20. Negovskii VA, Bel'skaia TP, Kassil' VL. [Work experience of a reanimation department.] *Sov Zdravookhr.* 1968;27:20–25.
21. Negovskii VA, Trusov LN, Ivanov EP. The pathogenesis of thrombo-haemorrhagic syndrome in agonal states and clinical death. *Cor Vasa.* 1968;10:165–172.
22. Safar P. Vladimir A. Negovsky, the father of "reanimatology." *Resuscitation.* 2001;49:223–229.
23. Behringer W. [Peter Safar—"father of resuscitation."] *Wien Klin Wochenschr.* 2004;116:102–106.
24. Mitka M. Peter J. Safar, MD: "father of CPR," innovator, teacher, humanist. *JAMA.* 2003;289:2485–2486.
25. Safar P. On the future of reanimatology. *Acad Emerg Med.* 2000;7:75–89.
26. Safar P. On the history of modern resuscitation. *Crit Care Med.* 1996;24:S3–11.
27. Safar P. Mild hypothermia in resuscitation: a historical perspective. *Ann Emerg Med.* 2003;41:887–888.
28. Brat R, Suk M, Barta J et al. [Resuscitation of a patient with deep hypothermia using extracorporeal circulation.] *Rozhl Chir.* 2002;81:279–281.
29. Cortes J, Galvan C, Sierra J, Franco A, Carceller J, Cid M. [Severe accidental hypothermia: rewarming by total cardiopulmonary bypass.] *Rev Esp Anestesiol Reanim.* 1994;41:109–112.
30. Karplus H. Suspended animation and resuscitation. A historical review in the light of experimental hypothermia. *J Forensic Med.* 1966;13:68–74.
31. Pinto D, Richard V. Suspended animation. Surg Forum 1962;13:191–193.
32. Machado C. Randomized clinical trial of magnesium, diazepam, or both after out-of-hospital cardiac arrest. *Neurology.* 2003;60:1868–1869.
33. Meyer JA. Claude Beck and cardiac resuscitation. *Ann Thorac Surg.* 1988;45(1):103–105.
34. Zoll PM, Linenthal AJ, Norman LR, Paul MH, Gibson W. Use of external electric pacemaker in cardiac arrest. *J Am Med Assoc* 1955;159(15):1428–1431.

35. Bruner H, Hornicke H, Stoffregen J. [Respiration of poliomyelitis patients in the iron lung.] *Dtsch Med Wochenschr.* 1954;79:1538.

36. Edmonson JM. A recent museum acquisition: an "iron lung." *Bull Cleve Med Libr Assoc.* 1982;28:14–15.

37. Gros J. [Use of the iron lung in polio.] *Munch Med Wochenschr.* 1954;96:837–839.

38. Lasley WH. The Draeger iron lung. *Med Bull US Army Eur.* 1955;12:11–12.

39. Maxwell J. The iron lung: halfway technology or necessary step? *Milbank Q.* 1986;64:3–29.

40. Murray AJ. An improved iron lung. *Br Med J* 1956;1(4979):1361.

41. Wilson DJ. Braces, wheelchairs, and iron lungs: the paralyzed body and the machinery of rehabilitation in the polio epidemics. *J Med Humanit.* 2005;26:173–190.

42. Worden G. Steel knives and iron lungs: medical instruments as medical history. *Caduceus.* 1993;9:111–118.

43. Wyatt HV. Iron lungs in the service of polio cases overseas in World War II. *J R Army Med Corps.* 1997;143:53–60.

44. Bing R, Gibbon JH Jr. Cardiopulmonary bypass—triumph of perseverance and character. *Clin Cardiol.* 1994;17:456–457.

45. Edmunds LH Jr. Advances in the heart-lung machine after John and Mary Gibbon. *Ann Thorac Surg.* 2003;76:S2220–S2223.

46. Settergren G. Brain death: an important paradigm shift in the 20th century. *Acta Anaesthesiol Scand.* 2003;47:1053–1058.

47. Leyden E. Beiträge und Untersuchungen zur Physiologie und Pathologie des Gehirns. *Virchows Arch* 1866;37:519–559.

48. Fagge H. In: Ossler W, ed. *Principles and Practise of Medicine*, 2nd edn. London: Macmillan, 1886:526,551.

49. Jalland. Cerebral abscess secondary to ear disease; trephining; death. *The Lancet* 1892; 1:527.

50. Macewen W. Symptoms of abscess of brain. In: *Pyogenic Infective Diseases of the Brain and Spinal Cord.* Glasgow: J Maclehose & Sons, 1893:196.

51. Horsley V. On the mode of death in cerebral compression and its prevention. *Q Med J.* 1894;2:306–309.

52. Duckworth D. Some cases of cerebral disease in which the function of respiration entirely ceases for some hours before that of the circulation. *Edinburgh Med J.* 1898;3:145–152.

53. Cushing H. Some experimental and clinical observations concerning states of increased intracranial tension. *Am J Med Sci.* 1902;124:375–400.

54. Cushing H. Physiologische und anatomische Beobachtungen über den Einfluss von Hirnkompression auf den intracraniellen Kreislauf und über einige hiermit verwandte Erscheinungen. *Mitteilungen Grenzgebieten Med Chirurg.* 1902;9:703–808.

55. Berger H. On the electroencephalogram of man. *Electroencephalogr Clin Neurophysiol.* 1969;suppl 28:37.

56. Berger H. Über das Elektrenkephalogramm des Menschen. *Arch Psychiat Nervenkr.* 1933;102:555–574.

57. Berger H. Über das Elektrenkephalogramm des Menschen. Neunte Mitteilung. *Arch Psychiat Nervenkr.* 1934;102:538–557.

58. Fischgold H. Hans Berger and his time. *Beitr Neurochir.* 1967;14:7–11.

59. Gerhard UJ, Schonberg A, Blanz B. ["If Berger had survived the Second World War, he certainly would have been a candidate for the Nobel Prize." Hans Berger and the legend of the Nobel Prize.] *Fortschr Neurol Psychiatr.* 2005;73:156–160.

60. Gloor P. The work of Hans Berger. *Electroencephalogr Clin Neurophysiol.* 1969;27:649.
61. Gloor P. Hans Berger and the discovery of the electroencephalogram. *Electroencephalogr Clin Neurophysiol.* 1969;suppl:36.
62. Haas LF. Hans Berger (1873–1941), Richard Caton (1842–1926), and electroencephalography. *J Neurol Neurosurg Psychiatry.* 2003;74:9.
63. Wiedemann HR. Hans Berger (1873–1941). *Eur J Pediatr.* 1994;153:705.
64. Crile GW, Dolley DH. An experimental research into the resuscitation of dogs killed by anesthetics and asphyxia. *J Exp Med.* 1906;8:713.
65. Crile GW, Telkes M, Rowland AF. The physical nature of death. *Sci Am.* 1930;143:30–32.
66. Sugar O, Gerard RW. Anoxia and brain potentials. *J Neurophysiol* 1938;1:558–572.
67. Crafoord C. Dödsorsaken vid obturerande lungemboli [The cause of death by obstructing pulmonary embolism.] *Nordisk Med.* 1939;2:1043–1044.
68. Hill L. On cerebral anaemia and the effects which follow ligation of the cerebral arteries. *Philos Trans.* 1900;39:69–122.
69. Riishede J, Ethelberg S. Angiographic changes in sudden and severe herniation of the brainstem through tentorial incisura. *Arch Neurol Psychiatry.* 1953;70:399–409.
70. Löfstedt S, von Reis G. Intrakraniella lesioner med bilateralt upphävd kontrastpassage i a. carotis interna [Intracranial lesions with abolished passage of x-ray contrast through the internal carotid arteries.] *Opusc Med.* 1956;1:199–202.
71. Löfstedt S, von Reis G. Diminution or obstruction of blood flow in the internal carotid artery. *Opusc Med.* 1959;4:345–360.
72. Wertheimer P, Jouvet M, Descotes J. A propos du diagnostic de la mort du système nerveux dans les comas avec arrêt respiratoire traites par respiration artficielle. *Presse Med.* 1959;67:87–88.
73. Jouvet M. Diagnostic électro-sous-cortico-graphique de la mort du système nerveux central au cours de certains comas. *Electroencephalogr Clin Neurophysiol.* 1959;11(4):805–808.
74. Mollaret P, Goulon M. Le coma dépassé (mémoire préliminaire). *Rev Neurol (Paris).* 1959;101:3–15.
75. Machado C. The first organ transplant from a brain-dead donor. *Neurology.* 2005;64:1938–1942.
76. Schwab RS, Potts F, Bonazzi A. EEG as an aid in determining death in the presence of cardiac activity (ethical, legal, and medical aspects). *Electroencephalogr Clin Neurophysiol.* 1963;15:147–148.
77. Wijdicks EF. The neurologist and Harvard criteria for brain death. *Neurology.* 2003;61:970–976.
78. A definition of irreversible coma. Report of the Ad Hoc Committee of the Harvard Medical School to Examine the Definition of Brain Death. *JAMA.* 1968;205:337–340.
79. Gilder SSB. Twenty-second World Medical Assembly. *Br Med J.* 1968;3:493–494.
80. Ulmann E. Experimentelle Nierentransplantation. *Wien Klin Wochenschr.* 1902;11:281–285.
81. Jaboulay M. Greffe du reins au pli du coude par soudure arte. *Bull Lyon Med.* 1906;107:575.
82. Carrel A. La technique operatoire des anastomoses vasculaires et la transplantation des visceres. *Lyon Med.* 1902;98:859.

83. Carrel A, Guthrie CC. Anastomoses of blood vessels by the patching method and transplantation of the kidney. *JAMA*. 1906;47:1648.

84. Voronoy U. [Blocking the reticuloendothelial system in man in some forms of mercuric chloride intoxication and the transplantation of the cadaver kidney as a method of treatment for the anuria resulting from the intoxication]. *Siglo Med*. 1937;97: 296.

85. Hamilton DNH, Reid WA. Yu Yu Voronoy and the first human kidney allograft. *Surg Gynecol Obstet*. 1984;159:289.

86. Starzl TE. History of clinical transplantation. *World J Surg*. 2000;24:759–782.

87. Kuss R, Teinturier J, Milliez P. Quelques essais de greffe rein chez l'homme. *Mem Acad Chir*. 1951;77:755.

88. Dubost C, Oeconomos N, Nenna A, Milliez P. Resultats d'une tentative de greffe renale. *Bull Soc Med Hop Paris*. 1951;67:1372.

89. Servelle M, Soulie P, Rougeulle J. Greffe d'une rein de supplicie a une malade avec rein unique congenital, atteinte de nephrite chronique hypertensive azatemique. *Bull Soc Med Hop Paris*. 1951;67:99.

90. Michon L, Hamburger J, Oeconomos N, et al. Une tentative de transplantation renale chez l'homme: aspects medicaux et biologiques. *Presse Med*. 1953;61:1419.

91. Hume DM, Merrill JP, Miller BF, Thorn GW. Experiences with renal homotransplantation in the human: report of nine cases. *J Clin Invest*. 1955;34:327.

92. Merrill JP, Murray JE, Harrison JH, Guild WR. Successful homotransplantation of the human kidney between identical twins. *JAMA*. 1956;160:277–282.

93. Murray JE. The first successful organ transplants in man. *J Am Coll Surg*. 2005;200:5–9.

94. Merrill JP, Murray JE, Takacs FJ, Harger EB, Wilson RE, Dammin GJ. Successful transplantation of kidney from a human cadaver. *JAMA*. 1963;185:347–353.

95. Starzl TE, Marchioro TL, Vonkaulla KN, Hermann G, Brittain RS, Waddell WR. Homotransplantation of the liver in humans. *Surg Gynecol Obstet*. 1963;117:659–676.

96. Hardy JD, Webb WR, Dalton ML Jr, Walker GR Jr. Lung homotransplantation in man. *JAMA*. 1963;186:1065–1074.

97. Lillehei RC, Idezuki Y, Feemster JA, et al. Transplantation of stomach, intestine, and pancreas: experimental and clinical observations. *Surgery*. 1967;62:721–741.

98. Kelly WD, Lillehei RC, Merkel FK, Idezuki Y, Goetz FC. Allotransplantation of the pancreas and duodenum along with the kidney in diabetic nephropathy. *Surgery*. 1967;61:827–837.

99. Schwartz RS. Immunosuppression—back to the future. *World J Surg*. 2000;24:783–786.

100. Barnard CN. The operation. A human cardiac transplant: an interim report of a successful operation performed at Groote Schuur Hospital, Cape Town. *S Afr Med J*. 1967;41:1271–1274.

101. Barnard CN. Human cardiac transplantation. An evaluation of the first two operations performed at the Groote Schuur Hospital, Cape Town. *Am J Cardiol*. 1968;22:584–596.

102. Barnard CN. What we have learned about heart transplants. *J Thorac Cardiovasc Surg*. 1968;56:457–468.

103. Barnard C. Reflections on the first heart transplant. *S Afr Med J*. 1987;72:xix-xxx.

104. Barnard CN. The first heart transplant—background and circumstances. *S Afr Med J*. 1995;85:924, 926.

105. Hoffenberg R. Christiaan Barnard: his first transplants and their impact on concepts of death. *BMJ.* 2001;323:1478–1480.
106. Barnard CN. Comments on the first human-to-human heart transplant. 1993. *Cardiovasc J S Afr.* 2001;12:192–194.
107. Cooper DK. Christiaan Barnard and his contributions to heart transplantation. *J Heart Lung Transplant.* 2001;20:599–610.
108. Brink JG, Cooper DK. Heart transplantation: the contributions of Christiaan Barnard and the University of Cape Town/Groote Schuur Hospital. *World J Surg.* 2005;29:953–961.
109. Horwitz NH, Dunsmore RH. Some factors influencing the nonvisualization of the internal carotid artery by angiography. *J Neurosurg.* 1956;13:155–164.
110. Wertheimer P, de Rougemont, Descotes J, Jouvet M. [Angiographical data concerning the death of the brain during comas with respiratory arrest (so-called protracted coma).] *Lyon Chir.* 1960 Sep;56:641–648.

2
The First Organ Transplant from a Brain-Dead Donor

At the beginning of 1959, Wertheimer et al. wrote about the "death of the nervous system."[1] Later that year, Mollaret and Goulon[2] coined the term *coma dépassé*, to describe an irreversible state of both coma and apnea. Those were the first attempts to characterize the clinical and pathophysiological condition of mechanically ventilated patients who had suffered the loss of brain functions but whose heartbeat was preserved.

The history of human organ transplantation had begun in 1954, when Joseph Murray, later a Nobel Laureate, and his team carried out an organ transplant in man, taking a kidney from an identical twin.[3] In 1962, Murray performed the first successful cadaveric kidney transplant.[3] In 1963, Thomas Starzl achieved the first human liver transplant,[4] and, in the same year, James D. Hardy performed the first lung transplant.[5]

In those days, the surgical team dealing with a brain-dead donor brought the donor into the operating room with the recipient for the removal; the respirator was then stopped, and everyone waited for the donor's heart to cease to beat.[6] Technically, therefore, these donors were not "brain-dead" at the time of organ retrieval. Rather, they had been declared dead by classic cardiorespiratory criteria. An irreversible loss of brain functions had not been tested in these cases.[7]

It is commonly believed that the first set of criteria for brain death (BD) originated in 1968,[8,9] with the report issued by the Ad Hoc Committee of the Harvard Medical School.[10] Wijdicks[11] recently published a full historical description of the development of this document. Actually, 5 years before the Harvard criteria appeared, Guy Alexandre, a Belgian surgeon at the Catholic University of Louvain, introduced a set of BD criteria based on the description of coma *dépassé*[2] and, on June 3, 1963, carried out the first transplant in his country. When Alexandre and his team performed the transplant, they did not discontinue mechanical ventilation and wait for the donor's heart to stop beating. Theirs was the first transplantation ever to make use of a heart-beating, brain-dead donor.[7,12–14]

I first knew about Dr. Guy Alexandre from Dr. Christopher Pallis's lecture during the First International Symposium on Brain Death, held in Havana, in 1992.[15] To investigate this milestone event, I reviewed the little-known proceedings of a Ciba Symposium held in London, on March 9–11, 1966.[16–18] I also corresponded with

Alexandre; Jean-Paul Squifflet, acting chairman of the Service of Kidney and Pancreatic Transplants and Surgery of Endocrine Glands of Louvain University; and Sir Roy Calne, who provided important information and historical documents.

Alexandre's Personal History

Guy Alexandre (Fig. 2.1) obtained a fellowship in surgical research at Harvard University, under Murray's supervision, in 1961–1962. He worked at the Peter Bent Brigham Hospital, directed by Francis D. Moore. His first contact in Boston was with Calne, who had also done a fellowship on Moore's service, but who at that time was packing to return to England. Calne entrusted to Alexandre the dogs surviving from his experiments.[19–23] Alexandre also worked with Hitchings and Elion (Nobel Prize co-recipients) on various immunosuppressive experimental protocols.[12.13.20–22.24]

After his fellowship, Alexandre returned to Belgium and immediately initiated steps to make kidney transplantation feasible. On June 3, 1963, a patient with a severe head injury was brought to the emergency department of the Saint Pierre Hospital in Louvain, in profound coma.[12] In spite of active resuscitation procedures and the administration of vasopressors and other drugs, the patient showed the clinical picture of coma *dépassé*.[2]

At Alexandre's request, Jean Morelle, chairman of the Department of Surgery, "took the most important decision of his career,"[12] allowing the removal of a kidney from that heart-beating patient. The graft functioned immediately after implant,

FIGURE 2.1. Guy P. J. Alexandre, Belgian, born on July 4, 1934. (Photo provided by Dr. Alexandre.) (See also color insert.)

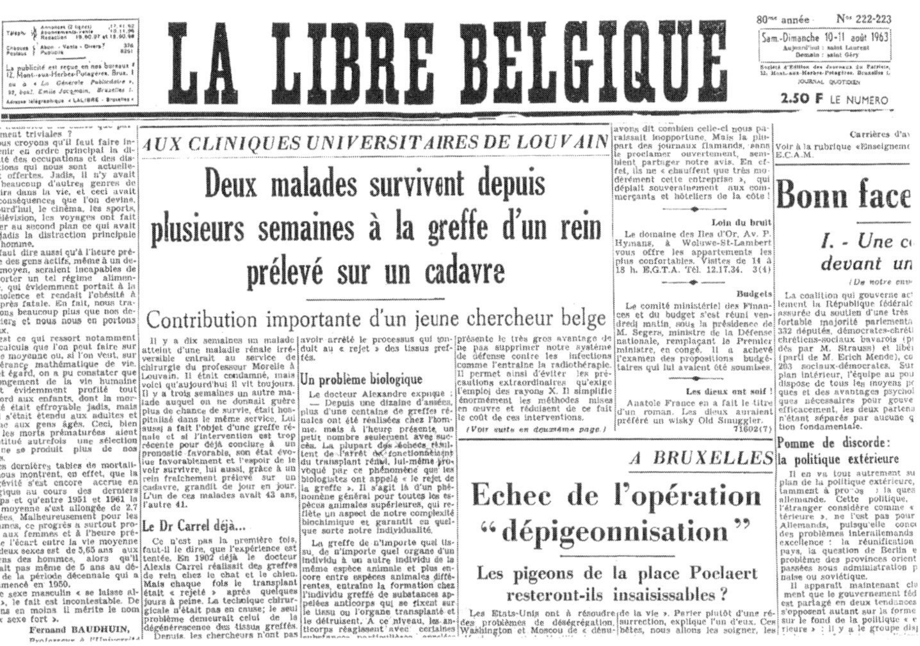

FIGURE 2.2. Article published in *La Libre Belgique*, August 10–11, 1963. The caption reads, "Two patients remain alive, several weeks after transplantation of a kidney previously removed from a cadaver. Important contribution of a young Belgian investigator." Provided by Dr. Guy P. J. Alexandre

without any sign of tubular necrosis, and serum levels of creatinine normalized in a few days. The recipient, who had been maintained by peritoneal dialysis, died of sepsis—with his new kidney in place—on day 87 (Fig. 2.2).[12,13]

The Ciba Symposium

The Ciba symposium on transplantation began to take shape when Michael Woodruff, from the Department of Surgical Science of the University of Edinburgh, pointed out to the Ciba Foundation's Director, Dr. Wolstenholme, "the need for a small conference of medical men, lawyers, and others concerned in the ethical and legal problems of organ transplantation."[25]

The conference had been organized originally without considering an invitation to Alexandre. Starzl had heard Alexandre's point of view concerning BD from Alexandre's colleague Otte, who was doing a year of specialized training in Starzl's department. Starzl suggested to the Foundation that Alexandre should be invited (Alexandre, personal communication) (Figs. 2.3 and 2.4).

From that symposium came a book that includes very important lectures and also records of discussions among the delegates.[25] Murray[26] presented a lecture, "Organ Transplantation: The Practical Possibilities." After this address, a discussion followed, in which Alexandre expressed his pioneering opinion (Fig. 2.2), remarking,

DOCTEUR GUY ALEXANDRE

BEVERLEE, January 30th, 1965
24, AVENUE LÉOPOLD III
TÉL. 016 - 206 76

Mrs Nancy Spufford
Conference Assistant
The Ciba Foundation
41,Portland Place
London W.1

Dear Mrs. Spufford,

          Thank you very much for your letter of
January 17th. I will be very happy to participate to the Symposiu
on"Ethics in Medical Progress", and would like to ask you
to thank Dr. Wolstenholme for me.
          I will certainly bring out in the discussions
our point of view in reference to the problem of cadaveric
transplantation as our position in that matter is quite different
from the other groups.
          Could you send me the application forms for
the travelling expenses as my funds for the time being are
very strictly limited. I would be very grateful if you could make
the necessary reservations for my wife and for me, from March 8
in the evening to March 11.
          I am joining to this letter a brief biographica
note, hoping that it will be satisfactory; please let me know
if you need a complete bibliography.
          I apologise for the late answer but I hope
it will not arrive to late. As a matter of fact, two months
ago, having heard of this Symposium, I wrote to the Medical
Department of Ciba who answered me that there was no Symposium
being planned on the Ethics of organ transplantation. You can imag
my surprise when I received your letter but I understand that
the invitations for this Conference are strictlylimited.
          Looking forward to further informations from
you, I remain,

                         Very sincerely yours,

                         Guy P.J.Alexandre,M.D.

FIGURE 2.3. Letter from Prof. Alexandre, dated January 30, 1965, to Mrs. Nancy Spufford, Conference Assistant for the Ciba Foundation. In the original document there is a mistake in date, because this letter was actually written in 1966 (Alexandre GPJ, personal communication). (Used with permission of Dr. Alexandre.)

To throw some fuel into the discussion, I would like to tell you what we consider as death when we have potential donors who have severe craniocerebral injuries. In nine cases we have used patients with head injuries, whose hearts had not stopped, to do kidney transplantations. Five conditions were always met in these nine cases: (1) complete bilateral mydriasis; (2) complete absence of reflexes, both natural and irresponsive to profound pain; (3) complete absence of spontaneous respiration, five minutes after mechanical respiration has been stopped; (4) falling blood pressure, necessitating increasing amounts of vasopressive drugs le 23 février 1966

T.E. STARZL, M.D.
Department of Surgery
Colorado General Hospital
DENVER (Colorado)

Dear Dr Starzl,

Thank you very much for having asked the Ciba
Foundation to invite me to the Symposium on Ethics Medical
Progress.

As you know, we have a different point of vieuw
in the utilization of cadaver organs and I hope this will be
discussed openly during the Symposium.

It is a great pity you cannot come in Brussel on
March 12th to preside the session on transplantation but I do
hope very much that you will be able to come in Louvain at
the occasion of your visit in Liège.

I think Dr J.B. OTTE is making good work in your
department and from what I hear from him, he is very happy
in his work, which is finally the most important.

I am looking forward to meet you in London and
hope again very much that you will be able to visit Louvain
in a near future.

Very sincerely yours,

Guy P.J.Alexandre,M.D.

FIGURE 2.4. Letter from Dr. Alexandre dated February 23, 1966, thanking Prof. Starzl for having proposed to the Ciba Foundation to invite him to the conference. (Used with permission of Dr. Alexandre.)

(either adrenaline or Neo-synephrine [phenylephrine hydrochloride]); (5) a flat EEG. All five conditions must be met before the removal of a kidney can be considered.

Alexandre, therefore, proposed one precondition—severe craniocerebral injury—and five criteria before BD could be diagnosed (Table 2.1). By applying these criteria, Alexandre and his team first diagnosed the donor as being dead. Afterward, the possible removal of the kidney was considered.[7]

The meaning of an isoelectric EEG was fully discussed in the Ciba Symposium. Hamburger, director of the Claude Bernard Institute of Research and Chief of the

TABLE 2.1. Alexandre's criteria[26] compared with Harvard criteria, as stated in the report[10] (see text)

Alexandre's Criteria	Harvard Criteria
Precondition: Severe craniocerebral injury	Precondition: Irreversible cerebral damage Exclusion of two conditions: hypothermia (below 90° F) and central nervous system depressants, such as barbiturates
Complete bilateral mydriasis	Pupil fixed and dilated and will not respond to a direct source of bright light.
Complete absence of reflexes, both natural and irresponsive to profound pain	Unreceptivity and unresponsivity to even the most intensely painful stimuli Ocular movement (to head-turning and to irrigation of the ears with ice water) and blinking are absent. There is no evidence of postural activity (decerebrate or other) Swallowing, yawning, vocalizations are in abeyance Corneal and pharyngeal reflexes are absent As a rule the stretch of [sic] tendon reflexes cannot be elicited Plantar or noxious stimulation gives no response
Complete absence of spontaneous respiration, five minutes after mechanical respiration has been stopped	No movements or breathing
Falling blood pressure, necessitating increasing amounts of vasopressor drugs: adrenaline or Neo-synephrine (phenylephrine hydrochloride)	Observations covering a period of at least 1 hour by physicians are adequate to satisfy the criteria of no spontaneous muscular movements or spontaneous respiration [established by turning off the respirator for 3 minutes] or response to stimuli such as pain, touch, sound, or light.
[<6 hours of observation if all five conditions are met]	All of the above tests shall be repeated at least 24 hours later with no change
Flat EEG	Flat EEG (when available it should be utilized)

Renal Unit at the Necker Hospital, in Paris, stated that he knew about two cases of coma due to severe barbiturate poisoning, with a flat EEG for several hours, followed by complete recovery.[27] He opined, "Those patients do not fulfill the other four conditions, nor do they have craniocerebral injuries; which I think is a very different situation."[26]

Murray then asked Alexandre: "Would you accept a flat EEG for four to six hours, along with your other four conditions, as incontrovertible evidence of death?"

Alexandre answered: "Using those five conditions you could not wait six hours, because falling blood pressure is one of the main conditions and after six hours the patients would already be dead anyway (conventionally dead). All our nine patients had blood pressures below 80 mm Hg after 30 or 45 minutes. They had had nearly half a litre of saline containing sometimes more than 160 mg noradrenaline to keep the blood pressure up."

Alexandre considered that the fall in blood pressure was irreversible in his cases, and denied that a patient in this state could be maintained for 6 hours. In recent years, it has been shown that, in some instances, human beings may have a long clinical existence after a declaration of BD, with preserved mechanisms for blood pressure control.[28,29]

He added, "The only way of not cheating the potential cadaver donor is to have two separate teams, one working to resuscitate the patient and the other taking care of the transplantation."[26] The Harvard committee later defended this point of view.[8,9,11]

Revillard (from Lyon, France) then argued, "We look for those five signs, Dr. Alexandre, and two others: (1) interruption of blood flow in the brain as judged by angiography, which we assume is a better sign of death than a flat EEG, and (2)—of less value—the absence of reaction to atropine. We do not quite agree with you on the falling blood pressure because the fall may often occur later than the other signs, and in some cases the blood pressure remains at 100 mm for several hours."[26]

At that time, Revillard already considered an absence of cerebral blood flow (CBF) to be a more powerful sign of death than a flat EEG. Alexandre reported that, in two of his cases, an angiogram showed that "the brain was not irrigated."[26] Bernat[30] has recently commented that the most certain way to show that the global loss of clinical brain functions is irreversible is to demonstrate the complete absence of CBF.

Hamburger then joined the discussion, adding a remarkable pathological explanation for Alexandre's proposal[31]:

It must be emphasized that this new approach to the definition of death has a serious pathological basis. We published in 1959 the case of a woman who had artificially maintained circulation and respiration but who otherwise fulfilled the criteria described by Dr. Alexandre and others... Dr. Ivan Bertrand, who is one of our best pathologists of the nervous system, did the autopsy: the entire nervous system including the brain and spinal cord had the appearance of a nervous system when the autopsy is done one week after death.[26]

In spite of Alexandre's advanced views, supported by Revillard and Hamburger, among others, serious doubts remained among the symposium attendees. Starzl and Calne, world leaders in organ transplantation at that time, expressed their doubts about accepting Alexandre's diagnostic criteria.[14,32,33] Starzl opined,

I doubt if any of the members of our transplantation team could accept a person as being dead as long as there was a heart beat. We have been discussing this practice in relation to renal homografts. Here, a mistake in evaluation of the "living cadaver" might not necessarily lead to an avoidable death since one kidney could be left. But what if the liver or heart were removed? Would any physician be willing to remove an unpaired vital organ before circulation had stopped?[26,33]

Calne said, "Although Dr. Alexandre's criteria are medically persuasive, according to traditional definitions of death he is in fact removing kidneys from live donors. I feel that if a patient has a heart beat he cannot be regarded as a cadaver."[34,35]

In a recent letter, Prof. Calne wrote,

I do very well remember the Ciba meeting on brain death in 1966. The main contribution of the meeting for me was Guy Alexandre's concept of coma dépassé. He pointed out that these criteria were used widely by neurologists as an indicator that further resuscitation and ventilation were fruitless and it was kindest to the relatives at this stage to tell them that with their consent ventilation would be stopped. Alexandre was, I think, the first to link this to the needs for transplantation surgery.

Starzl recently remarked,

At first, this idea appalled me because I envisioned that the care of a trauma victim could be jeopardized by virtue of his or her candidacy to become an organ donor. These fears were unfounded. The chances of a seriously injured patient being properly cared for were actually greatly increased when death was defined by the disappearance of brain function rather than the criteria of cessation of heart beat and respiration.[33]

The next paragraphs are quoted from Alexandre's recent messages to me:

I always thought that the way cadaver transplantation was performed in these early days in the few centers that performed them was totally hypocritical. As you know, they brought the donor (then called "coma dépassé") in the operating room with the recipient. They prepared the donor for the removal of the organ and THEN stopped the respirator and waited for the heart to stop. My point of view was that if the donor was already dead there was no need to damage the organ to be implanted by waiting for the heart to stop and to submit this organ to further ischemic damage; if he was not dead and by so doing, the last chance of survival of the donor was taken off.

The five criteria we used evolved from the many discussions I had with my colleagues. These included of course the neurologists and the colleagues of our intensive care units. These were, fortunately, dynamic persons whose competence in the field of neurology did not suffer any critics. I also had the good fortune of having a chief of the Department of Surgery, Prof. Jean Morelle, who was competent in neurosurgery and well in favor of kidney transplantation and who was ready to accept the responsibility of the kidney removal from patients in coma dépassé.

I asked Alexandre if his center declared the donor dead before the kidney was removed. He answered, "It is self evident that removing a kidney from a heart beating cadaver and allowing the reanimation team to stop the respirator once the kidney is removed (which was what was being done) would not have been done unless the person was declared dead."

He also remarked, "At the end of the Ciba symposium meeting you referred to, the president of the meeting asked the participants to let him know those who were prepared to act the way we were doing and to accept our criteria of brain death. I was the only one to raise my hand; all the others did not."

Evolution of an Idea

As Wijdicks[11] summarized the situation before the Harvard committee's report, "There was little consensus and little published literature before 1968." In France, the clinical entity of brain death had been reported by Wertheimer et al.[1] and Mollaret and Goulon.[2] In some modified form these criteria were mentioned at a transplantation meeting.[7]

Alexandre made his important decision to use a brain-dead, heart-beating donor only 5 years after the first characterizations of BD.[1,2] Although his notion was supported by some Ciba Symposium delegates, most attendees rejected it. Nonetheless, Alexandre had provoked a crucial discussion of the issue.[25]

After the Harvard committee's report was published, BD was widely accepted.[11] It is apparent that differentiating a new definition of death from applying that definition to the saving of lives helped the definition to gain acceptance. Similarly, differentiating the responsibility for deciding, clinically, that BD had occurred from the tasks of transplantation added to the acceptability of the new idea.[7] As Moore[14] emphasized, "How rapidly thinking has changed!"

This historical review briefly indicates how a position that, at first, was morally suspicious became ethically acceptable in the course of a few years. Moreover, it documents Alexandre's bold, practical contribution to the global discussion of human death. Alexandre did, indeed,"throw some fuel into the discussion."

References

1. Wertheimer P, Jouvet M, Descotes J. A propos du diagnostic de la mort du système nerveux dans les comas avec arrêt respiratoire traites par respiration artficielle. *Presse Med.* 1959;67:87–88.
2. Mollaret P, Goulon M. Le coma dépassé (mémoire préliminaire). *Rev Neurol (Paris).* 1959;101:3–15.
3. Merrill JP, Murray JE, Takacs FJ, Harger EB, Wilson RE, Dammin GJ. Successful transplantation of kidney from a human cadaver. *JAMA.* 1963;185:347–353.
4. Starzl TE, Marchioro TL, Vonkaulla KN, Hermann G, Brittain RS, Waddell WR. Homotransplantation of the liver in humans. *Surg Gynecol Obstet.* 1963;117:659–676.
5. Hardy JD, Webb WR, Dalton ML Jr, Walker GR Jr. Lung homotransplantation in man. *JAMA.* 1963;186:1065–1074.
6. Settergren G. Brain death: an important paradigm shift in the 20th century. *Acta Anaesthesiol Scand.* 2003;47:1053–1058.
7. Machado C. The first organ transplant from a brain-dead donor. *Neurology.* 2005;64:1938–1942.
8. Landmark article August 5, 1968: A definition of irreversible coma. Report of the Ad Hoc Committee of the Harvard Medical School to examine the definition of brain death. *JAMA.* 1984;252:677–679.
9. van Till–d'Aulnis de Bourouill HA. Diagnosis of death in comatose patients under resuscitation treatment: a critical review of the Harvard report. *Am J Law Med.* 1976;2:1–40.

10. A definition of irreversible coma. Report of the Ad Hoc Committee of the Harvard Medical School to Examine the Definition of Brain Death. *JAMA.* 1968;205:337–340.

11. Wijdicks EF. The neurologist and Harvard criteria for brain death. *Neurology.* 2003;61:970–976.

12. Squifflet JP. The history of transplantation at the Catholic University of Louvain, Belgium 1963–2003. *Acta Chir Belg.* 2003;103(3 Spec No):10–20.

13. Alexandre GPJ. From the early days of human kidney allotransplantation to prospective xenotransplantation. In: Terasaki PI, ed. *History of Transplantation: Thirty-Five Collections.* Los Angeles: U.C.I. Tissue Typing Laboratory; 1991:337–348.

14. Moore FD. Changing minds about brains. *N Engl J Med.* 1970;282:48.

15. Machado C, García OD, Román JM, Parets J. Four years after the "First International Symposium on Brain Death" in Havana: Could a definitive conceptual re-approach be expected? In: Machado C, ed. *Brain Death (Proceedings of the Second International Symposium on Brain Death).* Amsterdam: Elsevier Science, B. V., 1995:1–9.

16. Pallis C. Brainstem death. In: Braakman R, ed. *Handbook of Clinical Neurology: Head Injury.* Amsterdam: Elsevier Science B.V., 1990:441–496.

17. Pallis C. Further thoughts on brainstem death. *Anaesth Intensive Care.* 1995;23:20–23.

18. Pallis C. Brain stem death—the evolution of a concept. *Med Leg J.* 1987;55(pt 2):84–107.

19. Smith GT, Calne RY, Murray JE, Dammin GJ. Anatomic observations on the renal vessels in man with reference to kidney transplantation. *Surg Gynecol Obstet.* 1962;115:682–688.

20. Calne RY, Alexandre GP, Murray JE. A study of the effects of drugs in prolonging survival of homologous renal transplants in dogs. *Ann N Y Acad Sci.* 1962;99:743–761.

21. Calne RY. Biological factors in tissue transplantation. *Postgrad Med J.* 1962;38:548–559.

22. Calne RY. Prospects of renal transplantation. *Br J Clin Pract.* 1962;16:316–323.

23. Calne RY. Renal transplantation. *Mod Trends Surg.* 1962;1:137–151.

24. Alexandre GPJ, Murray JE. Further studies of renal homotransplantation in dogs. treated by combined Imuran therapy. *Surg Forum.* 1962;13:64–68.

25. Wolstenholme GEW, O'Connor M. *Ethics in Medical Progress: With Special Reference to Transplantation.* Boston: Little, Brown, 1966.

26. Murray JE. Organ transplantation: the practical possibilities. In: Wolstenholme GEW, O'Connor M, eds. *Ethics in Medical Progress: With Special Reference to Transplantation.* Boston: Little, Brown, 1966.

27. Mollaret P. [Physio-pathological treatment of barbiturate coma.] *J Med.* 1963;50:689–703.

28. Shewmon DA. Spinal shock and brain death: somatic pathophysiological equivalence and implications for the integrative-unity rationale. *Spinal Cord.* 1999;37:313–324.

29. Shewmon DA. Chronic "brain death": meta-analysis and conceptual consequences. *Neurology.* 1998;51:1538–1545.

30. Bernat JL. On irreversibility as a prerequisite for brain death determination. *Adv Exp Med Biol.* 2004;550:161–167.

31. Bertrand I, Lhermitte F. [Massive necrosis of the central nervous system in a subject kept alive artificially.] Rev Neurol (Paris) 1959;101:101–115.

32. Starzl TE. History of clinical transplantation. *World J Surg* 2000;24:759–782.
33. Starzl TE. *The Puzzle People—Memoirs of a Transplant Surgeon*. Pittsburgh: University of Pittsburgh Press, 1992.
34. Murray JE. Organ transplants: a type of reconstructive surgery. *Can J Surg.* 1965;8:340–350.

3
Conceptual Approach to Human Death on Neurological Grounds

Introduction

Any full account of death should include three distinct elements: the definition of death, the criterion (anatomical substratum) of brain death, and the tests to prove that the criterion has been satisfied.[1–13] Undoubtedly, the term 'criterion' for referring to the anatomical substratum introduces confusion in this discussion, because protocols of tests (clinical and instrumental) for brain diagnosis are called 'diagnostic criteria' or 'sets of diagnostic criteria'. Therefore, I will use the term 'anatomical substratum' instead of criterion.[14,15]

During the last decades, three main brain-oriented formulations of death have been discussed: whole brain, brainstem death and higher brain standards (Fig. 3.1).[14,16,17] The *whole brain* criterion refers to the irreversible cessation of all intracranial structure functions.[10,18–26] It has been accepted by society mainly for practical reasons.[14] Physicians have constructed batteries of bedside tests (and of confirmatory laboratory procedures) to show that this criterion of death has been satisfied.[27–30,30–35] Until recently, whole brain strategists had not provided a conceptual framework to support specific criteria and tests.[29,36] Moreover, this view has not answered the key point question about the critical number and location of neurons, subserving the essential brain activities to execute the functioning of the *"organism as a whole."*[37–67]

The brainstem standard was adopted in several Commonwealth countries.[68–75] Pallis emphasized that the capacity for consciousness and respiration are two hallmarks of life of the human being, and that brainstem death predicts an inescapable asystole.[76–78] However, a physiopathological review of consciousness generation will provide a basis for not accepting Pallis' definition of death.[14,15,79] Moreover, recent clinical cases have shown that brain death does not always predict an 'inevitable asystole within a short while'.[25,80–83]

The higher brain formulation springs largely from consideration of the persistent vegetative state (PVS), and has been mainly defended by philosophers.[14] The higher brain theorists have defined human death as the 'the loss of consciousness' (definition), related to the irreversible destruction of the neocortex (anatomical substratum).[18,22,26,84–99]

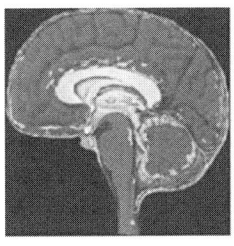 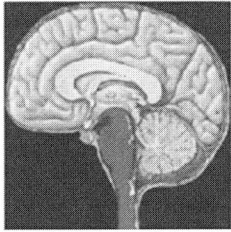 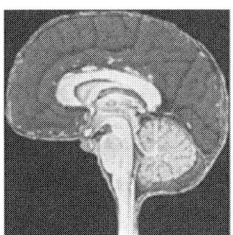

Whole brain   Brainstem death   Higher brain

FIGURE 3.1. Three main brain-oriented formulations of death have been discussed: whole brain, brainstem death, and higher brain standards. (See also color insert.)

I will demonstrate in this chapter that consciousness does not bear a simple one-to-one relationship with higher or lower brain structures and that, consequently, the higher brain view is wrong, because the definition (consciousness) does not harmonize with the anatomical substratum (neocortex). I will also discuss key aspects of the three brain-oriented formulations, and will propose a new standard of human death, based on the physiopathological mechanisms of consciousness generation.[14,16,79]

Mechanisms of Consciousness Generation

The term "consciousness" was defined by William James in 1890 as awareness of the self and the environment: *"Normal consciousness entails a state of being awake and aware of self and environment* (Figs 3.2A and 3.2B)."[100–103]

Plum and Posner also provided a similar definition of consciousness: *"the state of awareness of self and the environment (Fig 3.3)."*[104]

Two physiological components control conscious behavior: arousal and awareness (Fig. 3.4).[14–16,79,104,105] Arousal represents a group of behavioral changes that occurs when a person awakens from sleep or transits to a state of alertness.[106] "Normal consciousness requires arousal, autonomic-vegetative brain function subserved by ascending stimuli from the pontine tegmentum, posterior hypothalamus and thalamus that activate wakefulness."[107] The most discernible change that occurs when waking is the eyes opening. Arousal is also known as capacity for consciousness.[14–16,76,78,79,108,131,267]

Awareness, also known as content of consciousness, represents the sum of cognitive and affective mental functions, and denotes the knowledge of one's existence, and the recognition of the internal and external worlds.[14–16,79,103,109–120] It has been argued that consciousness has two dimensions: wakefulness and awareness.[107,121] Awareness is the same as the content of consciousness.[14–16,79] Wakefulness is provided by the arousal.[14–16,79,106] Arousal is supported by several brainstem neuron populations that directly project to both thalamic and cortical neurons. Therefore, decline of either the brainstem or both cerebral hemispheres may cause reduced

(A)

(B)

FIGURE 3.2. (A) The term *consciousness* was defined by William James in 1890 as awareness of the self and the environment. (B) The principles of psychology is considered a milestone contribution in psychology and the study of consciousness. Courtesy of the National
Library of Medicine.

FIGURE 3.3. From right to left: Prof. Fred Plum, Prof. Calixto Machado, Prof. and Mrs. Jerome Posner, and Dr. Yazmina Ferrer (Dr. Machado's wife). (See also color insert.)

wakefulness. Reflexes are important in the assessment of the functional integrity of the brainstem. However, severe impairment of brainstem reflexes can coexist with intact function of the reticular activating system if the tegmentum of the rostral pons and mesencephalon are unimpaired.

Plum has recently defined not two but three components, subdividing the content of consciousness in two levels or components.[105] According to this author,

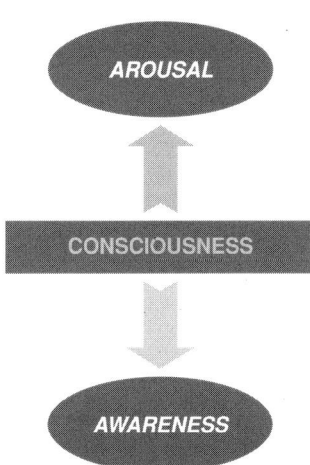

FIGURE 3.4. Two physiological components control conscious behavior: arousal and awareness.

the second component or level, "which importantly regulates the sustained behavioral state function of affect, mood, attention, cognitive integration, and psychic energy (cathexis) depends on the integrity of the limbic structures including the hypothalamus, the basal forebrain, the amygdala, the hippocampal complex, the cingulun, and the septal area." The limbic system is important for the homeostasis of the internal milieu, and hence the second component of consciousness is crucial for integrating affective, cognitive and vegetative functions. Plum considers the third component as the "cerebral level, along with the thalamus and basal ganglia." This component is related to the processes of higher levels of perception, self-awareness, language, motor skill, and planning. Memory can be impaired by injury of either cerebral or limbic levels.

In summary, a human being's state of consciousness reflects both his or her level of arousal that depends on subcortical arousal-energizing systems and, the sum of the cognitive, affective, and other higher brain functions (content of consciousness or awareness), related to "complex physical and psychologic mechanisms by which limbic systems and the cerebrum enrich and individualize human consciousness."[105] Therefore, I will use the term *arousal* when referring to those subcortical arousal-energizing systems, and *awareness*, to denote the sum of those complex brain functions, related to limbic and cerebrum levels.[14–16,79,104,105] Arousal disturbances lead to diminished alertness; content disturbances lead to diminished awareness and inattention, disorientation, and lack of integration of perception and processing memory.[105]

Unfortunately, most authors[17,122,123] mention human consciousness, without considering its two components originally described by Plum and Posner.[104] For example, higher brain theorists[18,22,26,84–86,89,90,93,99,122,124,125] habitually describe the persistent vegetative state (PVS) as patients with "irreversible loss of consciousness" or "permanent unconscious," but in these patients arousal is preserved, while awareness is apparently lost. On the other hand, some authors refer to the higher brain criterion as "the irreversible loss of the capacity for consciousness",[122] but they are really referring to awareness. As the use of the term "capacity for consciousness,"[20,76,108,126,127,267] could be confusing, I will identify this function with the original term used by Plum and Posner,[104] i.e., arousal. I will use awareness as a synonym for content of consciousness.[14–16,79]

Arousal depends on the integrity of physiological mechanisms that take their origin in the ascending reticular activating system (ARAS): "it originates in the upper brainstem reticular core and projects through synaptic relays in the thalamus to the cerebral cortex, where it increases excitability."[105] Moruzzi and Magoun,[128] in their pioneer studies, discovered "the presence in the brainstem of a system of ascending reticular relays, whose direct stimulation activates or desynchronizes the EEG, replacing high-voltage low waves with low voltage fast activity Figures 3.5A and 3.5B."

Nonetheless, Steriade et al.[129–142] have recently emphasized that this desynchronization related to wakefulness "is now more apparent than real," because although large slow waves disappear during waking, the EEG shows high frequency oscillations (30–40 Hz), known as gamma oscillations, that reflect synchronized and

(A)

Horace W. Magoun   Giuseppe Moruzzi

(B)

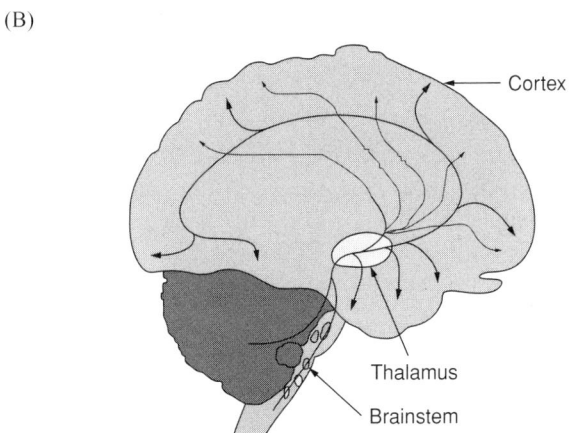

Thalamus

Brainstem

ASCENDING RETICULAR ACTIVATING SYSTEM

FIGURE 3.5. (A) Giuseppe Moruzzi and Horace W. Magoun discovered "the presence in the brainstem of a system of ascending reticular relays, whose direct stimulation activates or desynchronizes the EEG, replacing high-voltage low waves with low voltage fast activity" in 1949. (B) Ascending reticular activating system (ARAS). This system mainly provides arousal, and it is organized by several brainstem and diencephalic neuron populations that project to cortical neurons. Courtesy of the National Library of Medicine. (See also color insert.)

enhanced intracortical and corticothalamic activity. Singer and Gray have argued that these fast rhythms of corticothalamic neurons, known as gamma oscillations, are probably implicated in synchronizing mechanisms that respond to different features of the same perceptual object, leading to several hypotheses of high cognitive mechanisms.[143]

Bogen has emphasized that the intralaminar nuclei complex of the thalamus is a cardinal component of the ARAS.[144–146] The thalamic intralaminar neurons receive inputs from many sensory modalities and widely project to the cerebral cortex. Moreover, these nuclei are a major target for the brainstem reticular formation involved in waking. Recent reports strengthen the idea that intralaminar nuclei are thus essential in coordinating activity among cortical areas, and contribute to the formation of global perception to complex stimuli.[137,147–153]

Although initially the role of the mesencephalic reticular formation (MRF) and the thalamic intralaminar nuclei (ILN) was emphasized as simply mediating arousal, they are now considered as parts of integrating gating systems. Arousal is identified with the different functional states that characterize forebrain activation on the basis of brainstem modulation of corticothalamic systems; meanwhile, gating is formulated herein as a set of selective processes that may facilitate transient long-range interactions of large-scale brain networks.[110,140,154–156]

The connections from the brainstem to the cerebral cortex, relayed through intralaminar and other thalamic nuclei, and their main neurotransmitters (acetylcholine and glutamate) have been identified. These are neurotransmitter systems that take origin in the brainstem, hypothalamus and basal forebrain, projecting monosynaptically to the cerebral cortex without relaying through the thalamus. These systems include different neurotransmitter projections: cholinergic from the basal forebrain and mesopontine reticular formation, serotoninergic from the brainstem raphe nuclei, histaminergic from the posterior hypothalamus and noradrenergic from the brainstem locus coeruleus. Experimental studies have also shown that an almost complete destruction of the thalamus does not block cortical activation.[131,134,141,142] Furthermore, the EEG arousal pattern characterized by desynchronization disappears with the administration of drugs to block serotoninergic and cholinergic transmission.[157,158] Therefore, it is reasonable that arousal is due to several ascending systems stimulating the cerebral cortex and thalamus in parallel.[15,134,135,139,156,159,160] Thus, "thalamo-cortical transmission may not be sufficient or even necessary to produce cortical activation."[104] Moreover, the discovery of novel biologically active peptides has led to an explosion in our understanding of the molecular mechanisms that underlie the regulation of arousal, sleep and wakefulness, such as the neuropeptide S, urotensin II, the hypocretins, etc.[160–165]

Awareness is thought to be dependent upon the functional integrity of the cerebral cortex and its subcortical connections; each of its many parts are located, to some extent, in anatomically defined regions of the brain.[142,160,166,167] The discovery that the cerebral cortex is organized in vertical columns that represent functional units was crucial for further understanding of the functional organization of the brain. "The basic functional unit of the neocortex is a vertically oriented group of cells extending across the cellular layers and heavily interconnected in the vertical direction, sparsely so horizontally (Fig. 3.6)."[172] At present there are arguments that the functional organization of the entire cerebral cortex is a complex of these vertical columns. Contiguous columns are interconnected by local circuits into "information-processing modules," characterized by specific

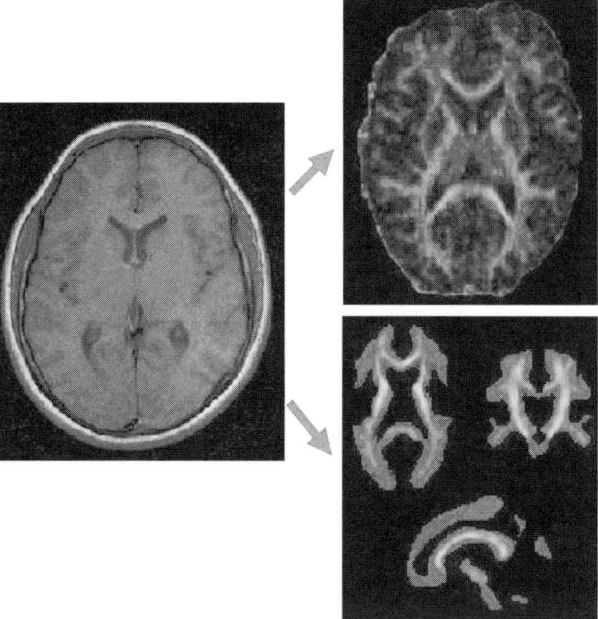

FIGURE 3.6. Cortical columns of the cerebral cortex. Awareness is thought to be dependent on the functional integrity of the cerebral cortex and its subcortical connections. (See also color insert.)

afferent and efferent connections with other modular units from other cortical and subcortical areas.[106,168,170]

According to Llinas, cognitive events depend on activity involving thalamocortical resonant colums. This author concluded that consciousness is a non-continuous event determined by synchronous activity in the thalamocortical system. Then, this synchronism could come about by the concurrent summation of specific and non-specific 40Hz activity along the radial dendritic axis of given cortical elements; that is, by coincidence detection. The specific system would provide the content in relation to information arising from the external world, and the non-specific system would endow the temporal conjunction, or the context, that would together engender a single cognitive experience.[151,171–187]

Singer and Gray have argued that fast rhythms of corticothalamic neurons, known as gamma oscillations, are probably implicated in synchronizing mechanisms that respond to different features of the same perceptual object, leading to several hypotheses of high cognitive mechanisms.[143]

Therefore, the cerebral cortex and thalamus make up "a unified oscillatory machine" that exhibit spontaneous rhythms and that are conditional to behavioral state and vigilance.[135] The brain uses spatiotemporally distributed systems to "capture high-order perceptual features."[188]

It seems that the brain operates in "parallel processing," because cortical regions are linked in networks parallel to each other and with subcortical structures. Thus, a specific component of a certain cognitive function is scattered among interconnected regions, each one implicated in a distinct aspect of the cognitive ability.[106,189–192] According to Feinberg, one of the most remarkable peculiarities of the brain is "the seemingly enormous redundancy, parallelism, and distributiveness" of its connections.[192] Figure 3.6 consciousness generation is based on substantial interconnections among the brainstem, subcortical structures and the cerebral cortex.

Additionally, Shewmon has discussed some examples of clear participation of subcortical structures in awareness.[193] Experimental animals with complete decortication have been shown to be capable of complex interactions with the environment, which is evidence of some awareness.[194] In lesions of the somatosensory cortex an evident loss of tactile, vibration and joint position sense is observed; nonetheless, conscious experience of pain and temperature is preserved, mediated by subcortical structures, probably the thalamus.[157,193] This author also commented that two hydranencephalic patients ("prenatal destruction of the cerebral hemispheres with intact skull and scalp") unquestionably manifested conscious behavior. These two cases are examples of the brainstem "plasticity" in newborns.[193,195] Clinical and experimental evidence convincingly suggests that the brainstem of newborns is potentially capable of much more complex integrative functioning. This includes some functions commonly considered to be cortical, even in animals.[195] Based on these subjects, the potential presence of some primitive form of awareness in anencephalics, and the possibility of subjective feeling of pain, has been suggested.[193,195,196] Thus, according to Shewmon "the human brainstem and diencephalon, in the absence of cerebral cortex, can mediate consciousness and purposeful interaction with the environment."[193]

The use of deep brain stimulation (DBS) has shown that is possible that the cerebral hemispheres could mediate arousal producing some wakefulness behavior, even after complete loss of the brainstem's reticular activating system.[14] Hassler et al. used DBS in "apallic" or "coma vigil" cases (PVS patients), stimulating the reticular formation in the thalamus and in the pallidum. It caused these patients to awaken with an undoubted recovery of awareness (recognition of their families and emotional expressions).[197,198] Sturm et al. reported the use of DBS at the thalamic level, in a case with probable dysfunction of the mesencephalic reticular formation due to the rupture of a sacular aneurysm at the tip of the basilar artery. DBS resulted in autonomic and behavioral reactions and the patient was able to respond to simple commands.[199] Kohadon et al., from a series of 25 PVS cases treated by DBS, reported a definitive improvement in arousal with some degree of awareness and interpersonal relationship, in 13 of them.[200,201] Katayama et al. also applied DBS in the mesencephalic reticular formation and/or thalamic center median-parafascicular (CM-pf) complex in PVS cases, with a follow-up of patients during 10 years. They reported that 8 of the 21 patients emerged from the VS, and became able to obey verbal commands, although they remained confined to a bed state except for one case. When they applied DBS in MCS, 4 of 5 cases emerged from being in a bedridden state, and were able to enjoy their life in their own home.[202–206,207,296] DBS therapy may be useful for allowing patients to emerge from the VS, if the candidates are selected according to appropriate neurophysiological criteria.[200,203,204,296] These authors also reported a persistent increment in pain-related P250, which indicates non-specific cortical activation.[207] They concluded that DBS therapy could be a useful tool for allowing patients to emerge from a VS, and to induce in MCS patients a consistent discernible behavioral evidence of consciousness recovery, and an emergence from confinement to bed.[200,202,204,206,207,208,296]

In spite of these reports about the use of DBS in PVS and MCS patients, Giacino considered that "deep-brain stimulation is an unproven treatment for facilitating recovery of consciousness in patients in VS," because according to this author, studies are characterized by significant deficiencies in methods.[209]

PVS provides an anatomic-functional model in which arousal is preserved and awareness is apparently lacking.[14,15,79,120,294] Therefore, it has been suggested that both components of consciousness "are mediated by distinct anatomic, neurochemical and/or physiological systems."[106] Nonetheless, the potential plasticity of the brain has demonstrated that subcortical structures could mediate awareness, even with the complete absence of the cerebral cortex.[82,83,195] Cases that have undergone hemispherectomy have shown clear signs of neuroplasticity.[210–214] Austin and Grant reported 3 cases that had undergone total hemispherectomy (comprising cortex, white matter and basal ganglia), who continued speaking and were aware of their environment during the operation, done under local anesthesia.[215] Recent reports remark that Anatomical hemispherectomy for intractable seizures provide an excellent seizure control, with low morbidity and no superficial brain haemosiderosis.[216]

Thus, awareness is not only related to the function of the neocortex (although it is of primary importance), but also to complex physical and psychological mechanisms, due to the interrelation of the ARAS, limbic system, and the cerebrum.[14,16,105,120,156,295]

Plum has emphasized that the ARAS substantially and inseparably activates and integrates both the arousal and the cognitive aspects of human consciousness. He recognized a brainstem-diencephalic participation not only in arousal, but also in cognitive function.[104,105,156,217] In lesions affecting thalamic-mesencephalic structures that comprise the ARAS, the presence of important cognitive and affective deficits can be found. Alterations in the cerebral cortex after severe damage restricted to mesencephalic-diencephalic activating systems have been reported. They reflect transneural degeneration, and suggest that these pathways not only activate the cerebral cortex but they also trophically influence cortical neurons.[105]

Therefore, it can be concluded that we cannot simply differentiate and locate arousal as a function of the ARAS, and awareness as a function of the cerebral cortex. Substantial interconnections among the brainstem, subcortical structures and the neocortex, are essential for subserving and integrating both components of human consciousness.[104,105,139,142,156,176,181,218,219]

The above considerations lead one to conclude that there is no single anatomical place of the brain "necessary and sufficient for consciousness."[193] Shewmon has discussed the existence of a "physiological kernel of consciousness" or a "reticular formation/cortical unit."[193,195,220] In a broad sense this "physiological kernel of consciousness" or "reticular formation/cortical unit" (RF/CU) is conformed by the widespread interconnections among the ARAS, diencephalon, other subcortical structures, and the cerebral cortex.[193]

Korein had emphasized that the brain is the critical system of the organism.[221–225] Nonetheless, Korein and I have recently presented a modification on the "Critical System of the Organism" concept, since we can further reduce the critical structure of the brain itself to the cortico-cerebral-thalamo-reticular complex. This "Critical System of the Brain" (CSB) complex includes the cerebral cortex, the basal ganglia, the thalamus, and the limbic system bilaterally, as well as the thalamic and brainstem reticular formations and components of the cerebellum. The anatomical and physiological substrate of the functions previously described requires the brainstem and thalamic reticular systems, the association nuclei of the thalamus and their connections. This critical system of the brain (CSB) is the minimal irreplaceable anatomical substrate of those functions (such as the multiple aspects of consciousness) that are utilized by the organism as a whole towards behavior that will result in decreased entropy production. Thus, onset of its development and function signify the onset of the life of the organism during the stage of its earliest appearance. Irreversible cessation of intrinsic CSB function indicates the death of the organism in the stage during which the CSB exists.[120]

Brain Oriented Standards of Death

To introduce the discussion about the three main standards of death on neurological grounds, I will consider the three distinct elements already mentioned: the definition of death, the anatomical substratum of brain death, and the tests to prove that the anatomical substratum is non-functioning.[1–5,7–9,226]

The Whole Brain Standard

James Bernat and his collaborators have presented the most complete defense of this standard.[1,3–5,9,10,12,13,226,227–236]

Definition

The permanent cessation of the functioning of the organism as a whole.

Early whole brain defenders[18–23,26–32,237–244] had not provided a conceptual framework to support this criterion until Bernat and his collaborators fully elaborated this formulation.[1,3–5,9,10,12,13,226,227–236] These authors defined death as "the permanent cessation of the functioning of the organism as a whole." By 'organism as a whole' they were not referring to the 'whole organism', as a sum of its parts, 'but rather to that characteristic that makes the living organism greater than the sum of its parts'. Furthermore, Bernat illustrated his conception of integration as follows: "functions of the organism as a whole include respiration, temperature control, fluid and electrolyte homeostasis, consciousness, food-seeking behavior, sexual behavior, neuroendocrine regulation, and autonomic control."[7] Bernat et al.

FIGURE 3.7. Whole brain standard defenders have defined death as "the permanent cessation of the functioning of the organism as a whole." (See also color insert.)

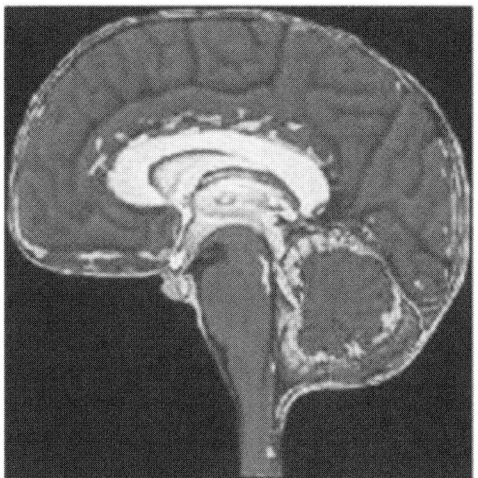

Whole Brain Standard

postulated that the organism as a whole could be still functioning, despite destruction of some subsystems.[1,8]

Anatomical Substratum

The permanent cessation of the functioning of the entire brain.
    The anatomical substratum of this view refers to the irreversible cessation of all intracranial structure functions.[1,3–5,9,10,12,13,226,227–236]

Tests

Bernat proposed two sets of tests: the permanent absence of breathing and heartbeat, and brain cessation tests. The cardiorespiratory tests are used to show the permanent loss of all brain functions, because a sustained arrest of circulation or respiration will produce ischemia, anoxia and subsequent necrosis of the brain. The cardiorespiratory tests are applied "in all cases except when death needs to be declared in a patient with heartbeat on a ventilator." Hence, 'brain death tests are necessary only for patients being mechanically ventilated.' These sets of criteria included a group of preconditions and a battery of tests and clinical procedures performed at the bedside. Periods of observation are required to carry out a second examination, in order to demonstrate the permanent absence of brain functions.[7]

Critique

Several authors have described patients as 'whole brain dead' and yet expressed surprise when they find that the EEG is retained. The persistence of EEG activity is indeed incompatible with the diagnosis of whole brain death.[67,245,246] The persistence of hypothalamic neuroendocrine functions in

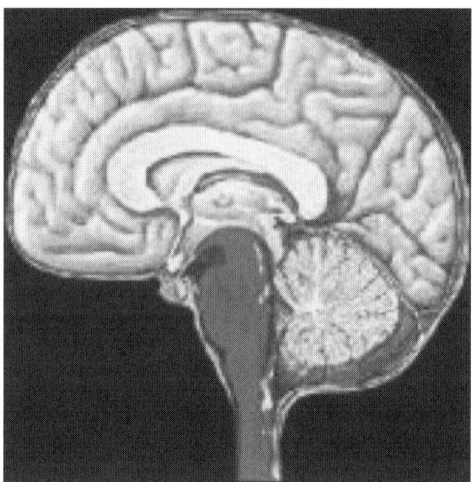

FIGURE 3.8. Brainstem death standard defenders have considered that the permanent cessation of the functioning of the brainstem means death. (See also color insert.)

Brainstem Death

otherwise whole brain-dead patients is likewise difficult to reconcile with 'whole brain death.[38,40–42,44,48,52,55,64,247–252]

These difficulties point to an underlying flaw in the 'whole-brain' account of death: it fails to specify the critical number and location of neurons required to subserve the essential activities of the hemispheres, diencephalon and brainstem, and thereby to execute the functions of the organism as a whole.[14–16,79]

The Brainstem Death Standard

Pallis powerfully defended this view both scientifically and philosophically. Moreover, this author was one of the first scholars trying to provide a definition of death on neurological grounds.[20,76–78,108,126,253–264] I will discuss this standard according to the three main elements: definition, anatomical substratum and tests.

Definition

There is only one kind of human death: the irreversible loss of the capacity for consciousness, combined with the irreversible loss of the capacity to breathe (and hence to sustain a spontaneous heartbeat).

Anatomical Substratum

The permanent cessation of the functioning of the brainstem.

Pallis postulated that the capacity for consciousness and respiration are two hallmarks of life for human beings, and that brainstem death predicts an inescapable asystole.[76,126,265] This author emphasized that the ascending reticular formation gives rise to a generalized activation of the cortex, producing the necessary arousal to endow the functioning of the "brain as a whole." He considered that the physiological and anatomical mechanism to abolish the capacity for consciousness is the irreversible damage of the paramedial tegmental areas of the mesencephalon and rostral pons.[76,108,127,254,256,265]

For many centuries, respiration was considered the crucial function that defined the frontiers between life and death.[266] In several ancient cultures death was considered: "the departure of the soul from the body," and hence the words that stand for soul are in many idioms the same as those standing for breath.[127,265]

Pallis also documented that the "loss of breath" or apnea relies on irreversible damage of the lower brainstem, where "crucial mechanisms concerned with breathing are located."[265]

This author presented a detailed review to answer the question: How long may cardiac action persist after a diagnosis of brain death? He emphasized that in most cases asystole occurred within days, and that time variations in somatic survival after brain death presumably reflect three main factors: (1) the time on the ventilator before the diagnosis of brain death was made; (2) the quality of life support administered; and (3) the age of the individual.[77,78]

Tests

According to Pallis "brainstem death is a clinical concept," and therefore "a dead brainstem" can be diagnosed at the bedside. The procedure is to diagnose an unconscious patient, with irreversible apnea and irreversible loss of brainstem reflexes, provided that all reversible causes of brainstem dysfunction have been excluded.[76,256]

Critique

Pallis includes in his definition "the capacity for consciousness" or arousal, as has been previously discussed.[77,265] Nonetheless, in any definition that incorporates consciousness as a main hallmark, both components should be included, because normal conscious behavior demands widespread interconnections among the ARAS, subcortical structures, and the neocortex, i.e., an interaction of both components.[14–16,79,104,105,120,156,217]

Moreover, some authors using deep brain stimulation have found nonspecific cortical activation in PVS and comatose patients.[48,197,198,202,204–206,207,296] Thus, in cases fulfilling the brainstem criteria of brain death with primary brainstem lesions and spared cerebral hemispheres, stimulation of the nonspecific thalamic nuclei might produce some degree of arousal that could endow awareness. This would surely refute the diagnosis of brain death.[14–16,79] In primary brainstem lesions a quasi-normal EEG could be recorded.[67,126,245,246,260,267,268,278,279]

Some recent reports have shown that some brain-dead patients do not develop an inevitable asystole within hours or days.[25,82,83,269–277] Shewmon has recently presented a detailed review of prolonged survivals in about 156 brain-dead patients.[81] This author compiled cases with survival of more than "a few days, i.e., one week or more," from sources including personal experience, the medical literature and the news media. He described the striking case of a patient with perhaps the longest

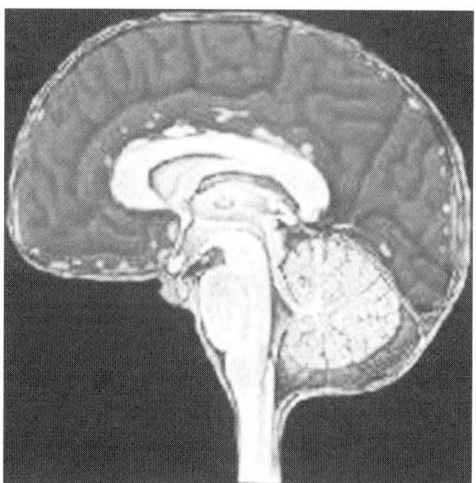

FIGURE 3.9. Higher brain advocates have proposed defining death as "the loss of that which is significant to the nature of man," which is related only to the cerebral cortex functioning. (See also color insert.)

Higher Brain Standard

recorded somatic survival, who was diagnosed (well documented) as 'brain-dead." He argued that the patient's age at onset of brain death plays a crucial role in somatic survival: 'the younger the age, the greater the capacity for survival'. This author also documented other factors implicated in survival: (1) 'associated systemic injuries directly due to whatever caused the brain insult'; and (2) 'systemic pathology secondarily induced by the process of brain herniation'. Withdrawal of life support is a 'confounding factor', because it can lead to an underestimate of the survival potential in brain-dead cases. The quality of nursing care, an adequate homeostatic control, prevention and early treatment of infections, etc., are other factors related to prolonged somatic survival. Consequently, an 'inevitable asystole' cannot be a justification for accepting a brain-oriented formulation of death.

The Higher Brain Standard (Figure 3.9)

Definition

The loss of that which is significant to the nature of humans.

'Higher brain' advocates proposed defining death as "the loss of that which is significant to the nature of man" and suggested that the irreversible loss of perception, sentience and cognition was necessary and sufficient for diagnosing death.[17,18,22,26,84-87,89,90,94,95,98,99,123,124,125,278-284] Bartlett and Younger asserted their belief "that only the higher brain functions, consciousness and cognition, define the life and death of a human being."[17]

Robert M. Veatch, pioneer of this standard of death, argued for including in the definition 'capacity for consciousness or social interaction', and emphasized the presence in human beings of 'the functions considered to be ultimately significant to human life': rationality, consciousness, personal identity, and social interaction.

This author proposed that death should be properly defined "as the irreversible loss of embodied capacity for social interaction."[285] Other authors have also proposed as the definition "the loss of personhood."[84,85]

Anatomical Substratum

The permanent cessation of the functioning of the neocortex.

Tests

No cognitive and affective functions.

The defenders of this formulation sustained that the neocortex assumes a critical role to provide consciousness and cognition. They functionally classified the brain into the lower brain (brainstem) that essentially controls vegetative functions, and the higher brain (the cerebral hemispheres, particularly the neocortex) that commands consciousness and cognition.[17,98,99,115,284–288] Veatch also referred to the "higher brain locus" or used other terms such as "cerebral," "cortical," or "neocortical." This author stated that "we could be quite conservative and hold that the entire brain must be destroyed in order to be sure that the capacity for consciousness and social interaction is lost." Veatch clearly argued that elaborating a set of tests to measure the irreversible loss of the capacity for consciousness or social interaction is rather difficult.[285]

Critique

As has been already discussed, arousal cannot simply be related to the function of the ARAS, and the content of consciousness related to the function of the cerebral cortex, because substantial interconnections among the brainstem, subcortical structures and the neocortex are indispensable for both components of human consciousness.[14–16,79,104,105,120,156,217] Hence, consciousness does not bear a simple one-to-one relationship with higher or lower brain structures, and the definition of consciousness does not correspond directly to the anatomical substratum (higher brain's anatomical locus). Unquestionably, the physical substratum for consciousness is based on anatomy and physiology throughout the brain.[14–16,79,120,159,289–290,291–295]

Higher brain advocates have stressed that PVS cases are dead.[17,84,86,99,122,123,284,285] The main finding in PVS is the preservation of arousal with apparent loss of awareness.[14,106,107,121,156,194,283,296–303,305,308,313] Kinney and Samuels emphasized that the PVS denotes a 'locked-out-syndrome', because the 'cerebral cortex is disconnected from the external world', explained by three main neuropathological patterns: widespread and bilateral lesions of the cerebral cortex, diffuse damage of intra- and subcortical connections in the white matter of the cerebral hemispheres, and necrosis of the thalamus.[106]

Can we deny the existence of internal awareness in PVS, because these patients apparently seem to be disconnected from the external world? The subjective dimension of awareness is philosophically impossible to test, but physiologically it seems conceivable that subjective awareness might continue.[295,294,304] Karen Ann

Quinlan's brain showed severe damage of the thalamus, with the cerebral hemispheres relatively spared,[298] and other authors have reported similar findings.[305–309] We can ask ourselves if, in a case like this, other activating pathways, projecting to the cerebral cortex without relaying through the thalamus, could stimulate the cerebral cortex to provide internal awareness, even if physicians are unable to detect its manifestations.[14]

Unexpected and well-documented recoveries of cognitive functions have been described in patients diagnosed by neurologists experienced and skilled in the diagnosis of this condition.[310–313] Childs et al. reported that 37% of 49 cases admitted to a special unit for rehabilitation were incorrectly diagnosed, according to the American Medical Association guidelines on PVS and the decision to withdraw support.[311] As it has been discussed, the use of DBS has been suggested as a potential method for recovery of awareness in PVS and functional recuperation in MCS cases.[202,207,296]

Thus, in PVS cases it is impossible to deny a possible preservation of internal awareness. According to the neuropathological pattern, either subcortical structures could provide internal awareness, or some remaining activating pathways projecting to the cerebral cortex without relaying through the thalamus could stimulate the cerebral cortex.[14–16,79,120] As consciousness is based on anatomy and physiology throughout the brain,[14–16,79,105,120,217] it is impossible to classify a PVS case as dead. The brain is severely damaged, but not fully and irreversibly destroyed.[106,298]

Moreover, it is necessary to consider the potential reversibility of awareness in the PVS, and an important recuperation in MCS. The development of rehabilitation techniques for PVS and MCS patients and others suffering long-lasting effects of brain injury is a crucial challenge for neuroscientists. The multisensory stimulation approach, the use of deep brain stimulation, the implementation of prosthetics to treat cognitive disabilities, the use of stem cells, and other techniques not yet developed, might provide new hope for cognitive rehabilitation of these cases.[197,198,200,202,204,206,207,296,302,311,313–325] This is another reason for not considering PVS patients as dead.

Thus, the main inconsistencies of the higher brain standard are:

(1) The definition of death does not correspond directly to the anatomical substratum, because proponents confuse the basis for consciousness with the neocortex.
(2) Proponents classify PVS cases as dead.

New Standard of Human Death

Definition Figure 3.10

Irreversible loss of the consciousness which provides the key human attributes and the highest level of control in the hierarchy of integrating functions within the human organism.

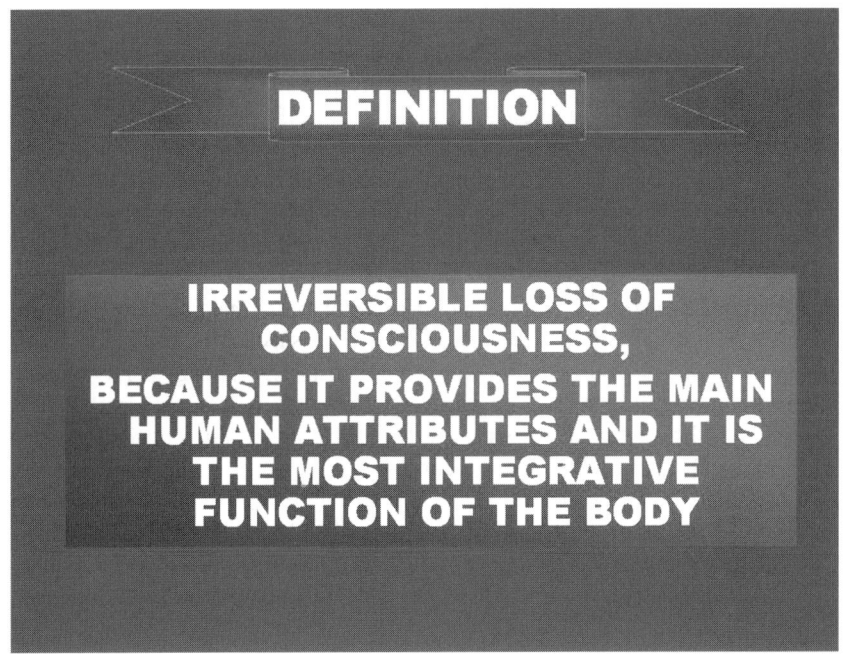

FIGURE 3.10. Definition of human death.

Botkin and Post drew an interesting distinction between major and minor clusters of attributes related to life. For example, brain dead patients retain several attributes associated with life, such as skin color, warm skin, heartbeat, kidney function, etc.[326] Even subjects who are dead according to the cardiorespiratory standard will preserve vestiges of life attributes for several days: hair and nails still grow.[266]

Cranford stated "our major premise is that consciousness is the most critical moral, legal, and constitutional standard, not for human life itself, but for human personhood."[99] Hence, higher brain defenders stressed that consciousness provides the most significant attributes of human existence.[17,18,22,26,84–87,90,93,94,98,99,123,124,278–282,284,285,327]

Therefore, the best candidate for the critical 'major attribute' of human life is consciousness: that is to say, without consciousness life loses most or all of its value for us. Consequently, I completely agree with higher brain advocates when considering that consciousness renders the most significant human attributes. Hence it is judicious to affirm that any vestige of consciousness is incompatible with death.[14–16,79,120,159,290,291,293–295,328,329]

Korein defended the notion of integration, applying thermodynamics and information theory to living systems. He documented that all living organisms may be classified as open systems that exchange energy and matter with their environment. This author maintained that any organism contains a critical system which 'supersedes all other subsidiary systems' or subsystems. Korein suggested that the 'critical system' of the human organism is the brain. According to his view, therefore, if the brain is irreversibly destroyed, the critical system is abolished. Even if other subsystems are functioning spontaneously or supported by machines, the organism as an individual entity no longer exists.[221,222,225,330,331] Korein and I have recently defended that the "Critical System of the Organism" is the cortico-cerebral-thalamo-reticular complex.

The Swedish Committee on Defining Death also defended the notion of integration, defining death as: "total and irreversible loss of all capacity for integrating and co-ordinating the functions of the body—physical and mental—into a functional unit."[332]

Shewmon has also been a defender of the central role of the brain "in the coordination or performance of virtually all functions necessary for the unity of the post-embryonic human body, including internal homeostasis, adaptive interaction with the environment, and the intimate connection between mental and physiological states."[193,194] However, this author recently changed his position (1997,1998a,b), when stating that the brain is not the 'central integrating organ of the body'.[80,81,194] He stated that 'clinical evidence of brain death is more attributable to multisystem damage and spinal shock than to destruction of the brain per se', and proposed to return to a 'circulatory respiratory' standard of death. However, this author accepted that the brain plays a role in integrating the functions of the intact organism. He used as an example the field of psychoneuroimmunology, emphasizing that "the brain role is one of modulating, fine tuning, and enhancing an already established and well functioning immune system."[82] If we accept Shewmon's view, then a specific emotional state could influence the immune system, either diminishing or enhancing the immune response. We can ask ourselves: Can we consider this brain's effect over other systems, of 'modulating' or 'fine-tuning', 'the highest level of integration within the organism'?[14–16,79]

Bernat affirmed that the brain generates signals "for breathing through brainstem ventilatory centers, and aids in the control of circulation through medullary blood pressure control centers."[7] For example, if a young woman sees a flower with its connotations of beauty and love, or sees an assassin with a knife trying to attack her, the brain develops a complex control mechanism over the whole body. After visualizing and recognizing (backup memory) the visual target (flower or assassin), complex signals are generated through pathways that interconnect extensive brain areas (neocortex, diencephalon and other limbic structures, brainstem, reticular formation, etc.) producing faster heartbeats and deeper ventilatory movements.[14,15,79] Trained Yoga practitioners are capable of slowing heartbeats until they are almost imperceptible to auscultation.[333–335] The psychological influences in the menstrual cycle, and in reproduction, have been extensively discussed.[336–338]

These examples demonstrate that consciousness controls and governs the functioning of the organism. This command over the organism can be conceived as 'modulating' or 'fine-tuning',[81,194,196] but it is, in fact, the highest level of control in the hierarchy of integrating functions within the human organism. Consciousness also stamps human individuality onto the integrated functioning of the body. Every subject responds in a different way to stimuli and daily life circumstances, according to his character, life experiences and personal interests.[15,16,79,120,292]

This concept of consciousness as the ultimate integrative function is more consistent with the biologically based system concepts of Korein,[221,222,224,237] and Bernat et al.[1,3,5,7,8,226] than the more philosophically based notions of personhood favoured by Veatch,[18,86,87,90,123,285] Barlett and Youngner,[17,84,85,339] and others.[98,99,122,340]

When I am explaining in a lecture my point of view about the role of consciousness in the organism, I emphasize that it is very common that a poet, a musician, etc., express: *"I love you from the depths of my heart."* Moreover, Valentine cards are always designed using hearts as symbols of love. Nothing is more distant to reality! We don't love with our hearts, but with our brains. Hence, in Figure 2.10a I try to illustrate that Cupid should send arrows to our brains and not to our hearts. Moreover, Valentine cards should be designed using brains and not hearts (Figure 2.10b), although I can't deny that by this way those cards would lose their enchantment (Figs 3.11A and 3.11B).

Anatomical Substratum Figure 3.12

Irreversible destruction of the anatomic and functional substratum of consciousness throughout the whole brain.

Substantial interconnections between the brainstem, subcortical structures and the neocortex are essential for subserving and integrating both components of human consciousness. Consequently, consciousness generation is based on anatomy and physiology throughout the brain.[104,105,128,131,134,156,218,298,308,341]

Tests Figure 3.13

Unresponsiveness, no arousal to any stimuli, and no cognitive and affective functions.

Implementing a reliable system to measure the irreversible loss of consciousness is quite difficult. The already difficult physiological evidence of loss of consciousness is complicated by the philosophical difficulty of assessing the subjective dimension of consciousness.[144,146,188,342] Most sets of tests measuring the absence of brain functions demand evidence of unresponsiveness. This is mainly explored by applying painful stimuli.[29,36,237] Painful stimuli explore the arousal component of consciousness that endows cognitive and affective functions.[16] Because the basis for consciousness is based on anatomy and physiology throughout the brain,[104,105,128,134,137,139,141,142,156,217,308,343,344] sets of diagnostic criteria should include tests to evaluate both brainstem and cerebral hemispheres.[79]

Comparison of the New Standard of Death with Other Standards Table 3.1

When this new standard of death is compared with whole brain, brainstem, and higher brain views, differences and similarities can be found.

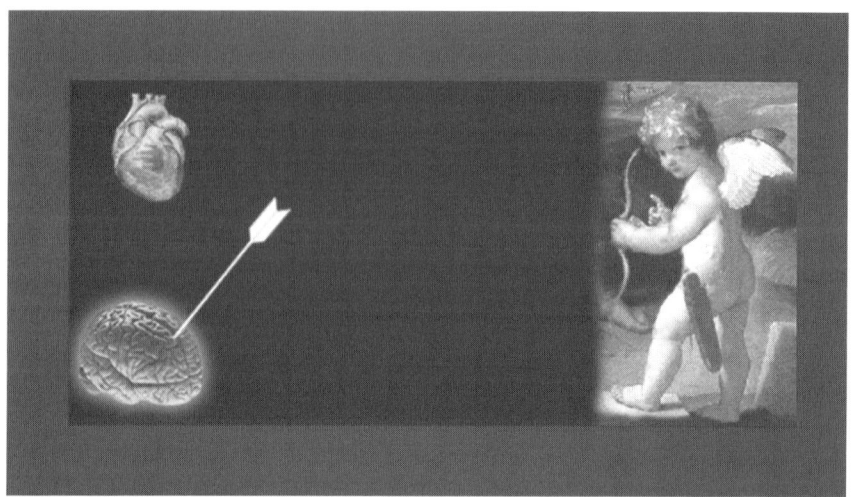

FIGURE 3.11. (A) Cupid shooting arrows into our brains. (B) Proposed Valentine's Day card.

Comparison with the Whole Brain Standard

Similarities: Both views are based on anatomy and physiology throughout the brain. PVS cases are classified as alive.

Differences: In my view, only one function (consciousness) is considered as the hallmark of the definition rather than all functions of the brain.

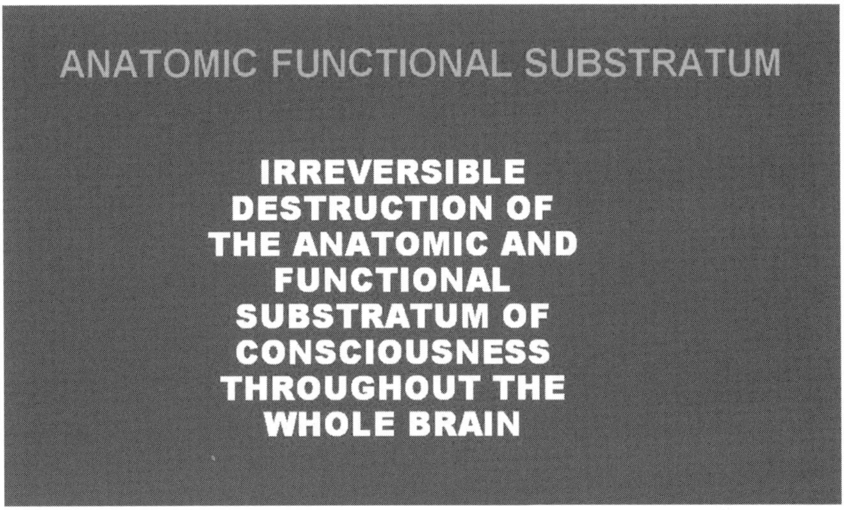

ANATOMIC FUNCTIONAL SUBSTRATUM

**IRREVERSIBLE
DESTRUCTION OF
THE ANATOMIC AND
FUNCTIONAL
SUBSTRATUM OF
CONSCIOUSNESS
THROUGHOUT THE
WHOLE BRAIN**

FIGURE 3.12. Anatomic substratum.

Comparison with the Brainstem Standard

Similarities: In both views one of the components of consciousness is included as a hallmark (capacity for consciousness or arousal). PVS cases are classified as alive.

Differences: In my view, both components of consciousness are included as hallmarks. In the brainstem view, the brainstem alone is considered as the anatomical substratum (criterion).

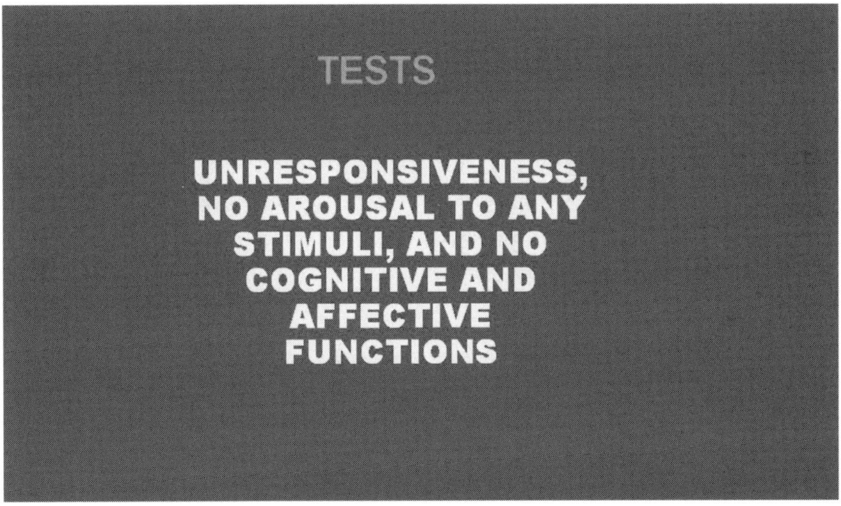

TESTS

**UNRESPONSIVENESS,
NO AROUSAL TO ANY
STIMULI, AND NO
COGNITIVE AND
AFFECTIVE
FUNCTIONS**

FIGURE 3.13. Tests.

TABLE 3.1. Comparison of the new standard of death with other standards

Standard	Definition	Anatomical substratum	Tests	Similarities with author's view	Differences from author's view
Whole brain	The permanent cessation of functioning of the organism as a whole	The permanent cessation of functioning of the entire brain	The permanent absence of breathing and heartbeat Brain cessation tests	Both views are based on anatomy and physiology throughout the brain; PVS cases are classified as alive	In my view, only one function (consciousness) is considered as the hallmark of the definition rather than all functions of the brain
Brainstem	There is only one kind of human death: the irreversible loss of the capacity for consciousness, combined with the irreversible loss of the capacity to breathe (and hence to sustain a spontaneous heart-beat)	The permanent cessation of functioning of brainstem	Irreversible apnea and irreversible loss of brainstem reflexes, provided that "all reversible causes of brainstem dysfunction have been excluded"	In both views, one of the components of consciousness is included as a hallmark (capacity for consciousness or arousal) PVS cases are classified as alive	In my view, both components of consciousness are included as hallmarks; In the brainstem view, the brainstem alone is considered as the anatomical substratum (criterion)
Higher brain	The loss of that which is significant to the nature of humans	The permanent cessation of the functioning of the neocortex	No cognitive and affective functions	In both views, consciousness is considered as the hallmark for defining death	In this view, the neocortex alone is considered as the anatomical substratum (criterion); PVS cases are classified as dead
New standard	Irreversible loss of the consciousness, which provides the key human attributes and the highest level of control in the hierarchy of integrating functions within the human organism	Irreversible destruction of the anatomic and functional substratum of consciousness throughout the whole brain	Unresponsiveness, no arousal to any stimuli, and no cognitive and affective functions		New definition identifies consciousness as the ultimate integrating function

PVS, persistent vegetative state.

Note: When this new standard of death is compared with whole brain, brainstem, and higher brain views, differences and similarities can be noted (see text for explanation).

Comparison with the Higher Brain Standard

Similarities: In both views consciousness is considered as the hallmark for defining death.

Differences: In this view, the neocortex alone is considered as the anatomical substratum. PVS cases are classified as dead. My definition identifies consciousness as the ultimate integrating function.

Conclusion

In conclusion, the best candidate for the critical 'major attribute' of human life is consciousness: that is to say, without consciousness life loses most or all of its value for us. Moreover, consciousness stamps human individuality on the integrated functioning of each of our bodies.

I therefore propose a standard of human death which identifies consciousness as the most important function of the body. I have emphasized that consciousness does not bear a simple one-to-one relationship with higher or lower brain structures, because the physical substratum for consciousness is based on anatomy and physiology throughout the brain.

References

1. Bernat JL, Culver CM, Gert B. On the definition and criterion of death. *Arch Intern Med* 1981:**94**: 389–394.
2. Bernat JL, Culver CM, Gert B. Definition of death. *Ann Intern Med* 1981:**95**: 652.
3. Bernat JL, Culver CM, Gert B. Defining death in theory and practice. *Hastings Cent Rep* 1982:**12**: 5–8.
4. Capron AM, Lynn J, Bernat JL, Culver CM, Gert B. Defining death: which way? *Hastings Cent Rep* 1982:**12**: 43–44.
5. Bernat JL. The definition, criterion, and statute of death. *Semin Neurol* 1984:**4**: 45–51.
6. Bernat JL, Culver CM, Gert B. Definition of death. *Ann Intern Med* 1984:**100**: 456.
7. Bernat JL. Ethical issues in neurology. Philadelphia: Lippincott,1991.
8. Bernat JL. Brain death. Occurs only with destruction of the cerebral hemispheres and the brain stem. *Arch Neurol* 1992:**49**: 569–570.
9. Bernat JL. How much of the brain must die in brain death? *J Clin Ethics* 1992:**3**: 21–26.
10. Bernat JL. A defense of the whole-brain concept of death. *Hastings Cent Rep* 1998:**28**: 14–23.
11. Bernat JL. Philosophical and Ethical Aspects of Brain Death. In: Wijdicks EFM (ed). *Brain Death*. Philadelphia: Lippincott Williams & Wilkins, 2001; 171–188.
12. Bernat JL. On irreversibility as a prerequisite for brain death determination. *Adv Exp Med Biol* 2004:**550**: 161–167.
13. Bernat JL. The concept and practice of brain death. *Prog Brain Res* 2005:**150**: 369–379.
14. Machado C. Death on neurological grounds. *J Neurosurg Sci* 1994:**38**: 209–222.
15. Machado C. Consciousness as a definition of death: its appeal and complexity. *Clin Electroencephalogr* 1999:**30**: 156–164.

16. Machado C. A new definition of death based on the basic mechanisms of consciousness generation in human beings. In: Machado C (ed). *Brain Death (Proceedings of the Second International Symposium on Brain Death)*. Amsterdam: Elsevier Science, BV, 1995; 57–66.
17. Bartlett ET, Youngner SJ. Human death and the destruction of the neocortex. In: Zaner RM (ed). *Death: Beyond the Whole-Brain Criteria*. New York: Kluwer Academic Publisher, 1988; 199–215.
18. Veatch RM. The whole-brain-oriented concept of death: an outmoded philosophical formulation. *J Thanatol* 1975:**3**: 13–30.
19. Walton DN. Neocortical versus whole-brain conceptions of personal death. *Omega (Westport)* 1981:**12**: 339–344.
20. Pallis C. Whole-brain death reconsidered—physiological facts and philosophy. *J Med Ethics* 1983:**9**: 32–37.
21. Browne A. Whole-brain death reconsidered. *J Med Ethics* 1983:**9**: 28–31, 44.
22. Veatch RM. The impending collapse of the whole-brain definition of death. *Hastings Cent Rep* 1993:**23**: 18–24.
23. Shann F. A personal comment: whole brain death versus cortical death. *Anaesth Intensive Care* 1995:**23**: 14–15.
24. Howsepian AA. In defense of whole-brain definitions of death. *Linacre Q* 1998:**65**: 39–61.
25. Potts M. A requiem for whole brain death: a response to D. Alan Shewmon's 'the brain and somatic integration'. *J Med Philos* 2001:**26**: 479–491.
26. Veatch RM. The death of whole-brain death: the plague of the disaggregators, somaticists, and mentalists. *J Med Philos* 2005:**30**: 353–378.
27. Walker AE. The death of a brain. *Johns Hopkins Med J* 1969:**124**: 190–201.
28. Smith AJ, Walker AE. Cerebral blood flow and brain metabolism as indicators of cerebral death. A review. *Johns Hopkins Med J* 1973:**133**: 107–119.
29. Walker AE, Molinari GF. Criteria of cerebral death. *Trans Am Neurol Assoc* 1975:**100**: 29–35.
30. Walker AE. Ancillary studies in the diagnosis of brain death. *Ann N Y Acad Sci* 1978:**315**: 228–240.
31. Walker AE. Advances in the determination of cerebral death. *Adv Neurol* 1979:**22**: 167–177.
32. Walker AE. Dead or alive. *J Nerv Ment Dis* 1984:**172**: 639–641.
33. Shiogai T, Saito I. [Ancillary studies in the determination of brain death]. *Nippon Rinsho* 1997:**55 Suppl 1**: 301–309.
34. Keske U. Tc-99m-HMPAO single photon emission computed tomography (SPECT) as an ancillary test in the diagnosis of brain death. *Intensive Care Med* 1998:**24**: 895–897.
35. Young GB, Lee D. A critique of ancillary tests for brain death. *Neurocrit Care* 2004:**1**: 499–508.
36. An appraisal of the criteria of cerebral death. A summary statement. A collaborative study. *JAMA* 1977:**237**: 982–986.
37. Dolinski A. Role of alanine in the production of endogenous glucose in patients with intracranial pathology. *Acta Neurochir Suppl (Wien)* 1979:**28**: 491–492.
38. Hall GM, Mashiter K, Lumley J, Robson JG. Hypothalamic-pituitary function in the "brain-dead" patient. *Lancet* 1980:**2**: 1259.
39. Schrader H, Krogness K, Aakvaag A, Sortland O, Purvis K. Changes of pituitary hormones in brain death. *Acta Neurochir (Wien)* 1980:**52**: 239–248.

40. Fiser DH, Jimenez JF, Wrape V, Woody R. Diabetes insipidus in children with brain death. *Crit Care Med* 1987:**15**: 551–553.
41. Keogh AM, Howlett TA, Perry L, Rees LH. Pituitary function in brain-stem dead organ donors: a prospective survey. *Transplant Proc* 1988:**20**: 729–730.
42. Arita K, Uozumi T, Oki S, Ohtani M, Taguchi H, Morio M. [Hypothalamic pituitary function in brain death patients—from blood pituitary hormones and hypothalamic hormones]. *No Shinkei Geka* 1988:**16**: 1163–1171.
43. Novitzky D, Cooper DK, Muchmore JS, Zuhdi N. Pituitary function in brain-dead patients. *Transplantation* 1989:**48**: 1078–1079.
44. Howlett TA, Keogh AM, Perry L, Touzel R, Rees LH. Anterior and posterior pituitary function in brain-stem-dead donors. A possible role for hormonal replacement therapy. *Transplantation* 1989:**47**: 828–834.
45. Hasegawa Y, Hasegawa T, Yokoyama T *et al.* Triphasic AVP secretion in encephalopathy. *Endocrinol Jpn* 1990:**37**: 171–175.
46. Fraser CL, Arieff AI. Fatal central diabetes mellitus and insipidus resulting from untreated hyponatremia: a new syndrome. *Ann Intern Med* 1990:**112**: 113–119.
47. Link J, Rohling R, Gramm HJ. [Preservation of homeostasis following the onset of brain death]. *Anaesthesiol Reanim* 1990:**15**: 249–260.
48. Yokota H, Nakazawa S, Shimura T, Kimura A, Yamamoto Y, Otsuka T. [Hypothalamic and pituitary function in brain death]. *Neurol Med Chir (Tokyo)* 1991:**31**: 881–886.
49. Harms J, Isemer FE, Kolenda H. Hormonal alteration and pituitary function during course of brain-stem death in potential organ donors. *Transplant Proc* 1991:**23**: 2614–2616.
50. Sazontseva IE, Kozlov IA, Moisuc YG, Ermolenko AE, Afonin VV, Ilnitskiy VV. Hormonal response to brain death. *Transplant Proc* 1991:**23**: 2467.
51. Gramm HJ, Meinhold H, Bickel U *et al.* Acute endocrine failure after brain death? *Transplantation* 1992:**54**: 851–857.
52. Sugimoto T, Sakano T, Kinoshita Y, Masui M, Yoshioka T. Morphological and functional alterations of the hypothalamic-pituitary system in brain death with long-term bodily living. *Acta Neurochir (Wien)* 1992:**115**: 31–36.
53. Galinanes M, Smolenski RT, Hearse DJ. Brain death-induced cardiac contractile dysfunction and long-term cardiac preservation. Rat heart studies of the effects of hypophysectomy. *Circulation* 1993:**88**: II270–II280.
54. Ujihira N, Hashizume Y, Takahashi A. A clinico-neuropathological study on brain death. *Nagoya J Med Sci* 1993:**56**: 89–99.
55. Arita K, Uozumi T, Oki S, Kurisu K, Ohtani M, Mikami T. The function of the hypothalamo-pituitary axis in brain dead patients. *Acta Neurochir (Wien)* 1993:**123**: 64–75.
56. Wang LC, Cohen ME, Duffner PK. Etiologies of central diabetes insipidus in children. *Pediatr Neurol* 1994:**11**: 273–277.
57. Nishioka H, Ito H, Ikeda Y, Koike S, Ito Y. [Hypothalamic-adenohypophysial function of corticotropin releasing hormone (CRH)-ACTH system in brain death]. *No To Shinkei* 1995:**47**: 1159–1163.
58. Bittner HB, Kendall SW, Chen EP, Van TP. Endocrine changes and metabolic responses in a validated canine brain death model. *J Crit Care* 1995:**10**: 56–63.
59. Bittner HB, Kendall SW, Campbell KA, Montine TJ, Van TP. A valid experimental brain death organ donor model. *J Heart Lung Transplant* 1995:**14**: 308–317.
60. Chen EP, Bittner HB, Kendall SW, Van TP. Hormonal and hemodynamic changes in a validated animal model of brain death. *Crit Care Med* 1996:**24**: 1352–1359.

61. Ponomarev AB. [Brain death caused by catecholamine crisis during adrenalectomy]. *Anesteziol Reanimatol* 1996: 76–79.
62. Hayashi S, Isayama K, Teramoto A. [Anterior pituitary functions in patients with severe head injuries treated with moderate hypothermia]. *No To Shinkei* 1996:**49**: 145–150.
63. Colpart JJ, Ramella S, Bret M *et al.* Hypophysis-thyroid axis disturbances in human brain-dead donors. *Transplant Proc* 1996:**28**: 171–172.
64. Nishioka H, Ito H, Ikeda Y, Koike S. [Hypothalamic-adenohypophysial function of corticotropin releasing hormone (CRH)-ACTH system in hypoxic coma]. *No To Shinkei* 1996:**48**: 53–57.
65. Wei B, Li D, Wang D. [Macro, micro and ultrastructural observations of pituitary, adrenal, pancreas and spleen in experimental brain death]. *Fa Yi Xue Za Zhi* 1997:**13**: 195–8, 253.
66. Lopau K, Mark J, Schramm L, Heidbreder E, Wanner C. Hormonal changes in brain death and immune activation in the donor. *Transpl Int* 2000:**13 Suppl 1**: S282–S285.
67. Ashwal S, Schneider S. Failure of electroencephalography to diagnose brain death in comatose children. *Ann Neurol* 1979:**6**: 512–517.
68. Diagnosis of brain death. Statement issued by the honorary secretary of the Conference of Medical Royal Colleges and their Faculties in the United Kingdom on 11 October 1976. *Br Med J* 1976:**2**: 1187–1188.
69. Diagnosis of death. Memorandum issued by the honorary secretary of the Conference of Medical Royal Colleges and their Faculties in the United Kingdom on 15 January 1979. *Br Med J* 1979:**1**: 332.
70. Criteria for the diagnosis of brain stem death. Review by a working group convened by the Royal College of Physicians and endorsed by the Conference of Medical Royal Colleges and their Faculties in the United Kingdom. *J R Coll Physicians Lond* 1995:**29**: 381–382.
71. Jennett B. Diagnosis of brain death. *J Med Ethics* 1977:**3**: 4–5.
72. Jennett B. Brain death 1981. *Scott Med J* 1981:**28**: 191–193.
73. Jennett B. Brain death. *Intensive Care Med* 1982:**8**: 1–3.
74. Jennett B. Brain death 1983. *Practitioner* 1983:**227**: 451–454.
75. Jennett B. Brain stem death defines death in law. *BMJ* 1999:**318**: 1755.
76. Pallis C. Brain stem death–the evolution of a concept. *Med Leg J* 1987:**55 (Pt 2)**: 84–107.
77. Pallis C. Brainstem death. *Handbook of Clinical Neurology: Head Injury*, Vol 13, Braakman R edn. Amsterdam: Elsevier Science Publisher BV, 1990; 441–496.
78. Pallis C. Further thoughts on brainstem death. *Anaesth Intensive Care* 1995:**23**: 20–23.
79. Machado C. Is the concept of brain death secure? In: Zeman A, Enamorado A (eds). *Ethical Dilemmas in Neurology*, Vol 36. London: W. B. Saunders Company, 2000; 193–212.
80. Shewmon DA. Recovery from "brain death": a neurologist's apologia. *Linacre Q* 1997:**64**: 30–96.
81. Shewmon DA. Chronic "brain death": meta-analysis and conceptual consequences. *Neurology* 1998:**51**: 1538–1545.
82. Shewmon DA. Spinal shock and 'brain death': somatic pathophysiological equivalence and implications for the integrative-unity rationale. *Spinal Cord* 1999:**37**: 313–324.
83. Shewmon AD. The brain and somatic integration: insights into the standard biological rationale for equating "brain death" with death. *J Med Philos* 2001:**26**: 457–478.

84. Green MB, Wikler D. Brain death and personal identity. *Philos Public Aff* 1980:**9**: 105–133.

85. Brody H. Brain death and personal existence: a reply to Green and Wikler. *J Med Philos* 1983:**8**: 187–196.

86. Veatch RM. Brain death: Welcome definition—or dangerous judgement? *Hastings Cent Rep* 1972:**2**: 10–13.

87. Veith FJ, Fein JM, Tendler MD, Veatch RM, Kleiman MA, Kalkines G. Brain death. I. A status report of medical and ethical considerations. *JAMA* 1977:**238**: 1651–1655.

88. Veatch RM. Death and dying: the legislative options. *Hastings Cent Rep* 1977:**7**: 5–8.

89. Veatch RM. The definition of death: ethical philosophical, and policy confusion. *Ann N Y Acad Sci* 1978:**315**: 307–321.

90. Veatch RM. Defining death: the role of brain function. *JAMA* 1979:**242**: 2001–2002.

91. Veatch RM, Tai E. Talking about death: patterns of lay and professional change. *Ann Am Acad Pol Soc Sci* 1980: 29–45.

92. Veatch RM. Comments on "Declaration of brain death in neurosurgical and neurological practice". *Neurosurgery* 1984:**15**: 457–458.

93. Veatch RM. Defining death at the beginning of life. *Second Opin* 1989: 51–59.

94. Veatch RM. Nancy Cruzan and the best interest standard. *Midwest Med Ethics* 1990:**6**: 17–19.

95. Veatch RM. Brain death and slippery slopes. *J Clin Ethics* 1992:**3**: 181–187.

96. Veatch RM. The dead donor rule: true by definition. *Am J Bioeth* 2003:**3**: 10–11.

97. Veatch RM. Abandon the dead donor rule or change the definition of death? *Kennedy Inst Ethics J* 2004:**14**: 261–276.

98. Cranford RE, Smith HL. Some critical distinctions between brain death and the persistent vegetative state. *Ethics Sci Med* 1979:**6**: 199–209.

99. Cranford RE, Smith DR. Consciousness: the most critical moral (constitutional) standard for human personhood. *Am J Law Med* 1987:**13**: 233–248.

100. James W. *The principles of psychology*, Vol One (originally published in 1890). New York: Dover,1950.

101. Lloyd D. Beyond "the fringe": A cautionary critique of William James. *Conscious Cogn* 2000:**9**: 629–637.

102. Chafe W. A linguist's perspective on William James and "The stream of thought". *Conscious Cogn* 2000:**9**: 618–628.

103. Epstein R. The neural-cognitive basis of the Jamesian stream of thought. *Conscious Cogn* 2000:**9**: 550–575.

104. Plum F, Posner JB. *The Diagnosis of Stupor and Coma*. Philadelphia: FA Davis, 1980.

105. Plum P. Coma and related global disturbances of the human conscious state. In: Peters A (ed). *Cerebral Cortex*, Vol 9. New York: Plenum, 1991; 359–425.

106. Kinney HC, Samuels MA. Neuropathology of the persistent vegetative state. A review. *J Neuropathol Exp Neurol* 1994:**53**: 548–558.

107. Medical aspects of the persistent vegetative state (1). The Multi-Society Task Force on PVS. *N Engl J Med* 1994:**330**: 1499–1508.

108. Pallis C, MacGillivary B. Brain death. *Lancet* 1981:**1**: 223.

109. Damasio AR. Consciousness. Knowing how, knowing where. *Nature* 1995:**375**: 106–107.

110. Damasio AR. Investigating the biology of consciousness. *Philos Trans R Soc Lond B Biol Sci* 1998:**353**: 1879–1882.

111. Damasio AR. How the brain creates the mind. *Sci Am* 1999:**281**: 112–117.

112. Parvizi J, Damasio A. Consciousness and the brainstem. *Cognition* 2001:**79**: 135–160.

113. McMahan J. The metaphysics of brain death. *Bioethics* 1995:**9**: 91–126.
114. Tononi G, Edelman GM. Consciousness and complexity. *Science* 1998:**282**: 1846–1851.
115. Lizza JP. Defining death for persons and human organisms. *Theor Med Bioeth* 1999:**20**: 439–453.
116. Taylor JG. The central role of the parietal lobes in consciousness. *Conscious Cogn* 2001:**10**: 379–417.
117. Thompson RF. Commentary on E. R. John et al. "Invariant reversible QEEG effects of anesthetics" and E. R. John "A field theory of consciousness". *Conscious Cogn* 2001:**10**: 245–258.
118. John ER. A field theory of consciousness. *Conscious Cogn* 2001:**10**: 184–213.
119. John ER. The neurophysics of consciousness. *Brain Res Brain Res Rev* 2002:**39**: 1–28.
120. Korein J, Machado C. Brain death: updating a valid concept for 2004. *Adv Exp Med Biol* 2004:**550**: 1–14.
121. Medical aspects of the persistent vegetative state (2). The Multi-Society Task Force on PVS. *N Engl J Med* 1994:**330**: 1572–1579.
122. Truog RD, Fackler JC. Rethinking brain death. *Crit Care Med* 1992:**20**: 1705–1713.
123. Veatch RM. The definition of death: ethical, philosophical, and policy confusion. In: Korein J (ed). *Brain Death: Interrelated Medical and Social Issues*, Vol 315. New York: Ann NY Acad Sci, 1977; 305–317.
124. Veith FJ, Fein JM, Tendler MD, Veatch RM, Kleiman MA, Kalkines G. Brain death. II. A status report of legal considerations. *JAMA* 1977:**238**: 1744–1748.
125. Veatch RM. What it means to be dead. *Hastings Cent Rep* 1982:**12**: 45–46.
126. Frackowiak R, Pallis C. Brain storm. *New Sci* 1980:**88**: 390.
127. Pallis C. Defining death. *Br Med J (Clin Res Ed)* 1985:**291**: 666–667.
128. Moruzzi G, Magoun HW. Brain stem reticular formation and activation of the EEG. *Electroencephalography and Clinical Neurophysiology* 1949:**1**: 455–473.
129. Nunez A, Amzica F, Steriade M. Voltage-dependent fast (20–40 Hz) oscillations in long-axoned neocortical neurons. *Neuroscience* 1992:**51**: 7–10.
130. Destexhe A, Contreras D, Sejnowski TJ, Steriade M. A model of spindle rhythmicity in the isolated thalamic reticular nucleus. *J Neurophysiol* 1994:**72**: 803–818.
131. Steriade M. Brain activation, then (1949) and now: coherent fast rhythms in corticothalamic networks. *Arch Ital Biol* 1995:**134**: 5–20.
132. Steriade M, Contreras D, Amzica F, Timofeev I. Synchronization of fast (30–40 Hz) spontaneous oscillations in intrathalamic and thalamocortical networks. *J Neurosci* 1996:**16**: 2788–2808.
133. Steriade M, Amzica F. Intracortical and corticothalamic coherency of fast spontaneous oscillations. *Proc Natl Acad Sci U S A* 1996:**93**: 2533–2538.
134. Steriade M. Arousal: revisiting the reticular activating system. *Science* 1996:**272**: 225–226.
135. Steriade M. Synchronized activities of coupled oscillators in the cerebral cortex and thalamus at different levels of vigilance. *Cereb Cortex* 1997:**7**: 583–604.
136. Timofeev I, Steriade M. Fast (mainly 30–100 Hz) oscillations in the cat cerebellothalamic pathway and their synchronization with cortical potentials. *J Physiol* 1997:**504 (Pt 1)**: 153–168.
137. Bazhenov M, Timofeev I, Steriade M, Sejnowski TJ. Cellular and network models for intrathalamic augmenting responses during 10-Hz stimulation. *J Neurophysiol* 1998:**79**: 2730–2748.

138. Steriade M, Timofeev I, Durmuller N, Grenier F. Dynamic properties of corticothalamic neurons and local cortical interneurons generating fast rhythmic (30–40 Hz) spike bursts. *J Neurophysiol* 1998:**79**: 483–490.

139. Steriade M. The corticothalamic system in sleep. *Front Biosci* 2003:**8**: d878–d899.

140. Steriade M. Local gating of information processing through the thalamus. *Neuron* 2004:**41**: 493–494.

141. Fuentealba P, Steriade M. The reticular nucleus revisited: intrinsic and network properties of a thalamic pacemaker. *Prog Neurobiol* 2005:**75**: 125–141.

142. Steriade M. Grouping of brain rhythms in corticothalamic systems. *Neuroscience* 2006:**137**: 1087–1106.

143. Singer W, Gray CM. Visual feature integration and the temporal correlation hypothesis. *Annu Rev Neurosci* 1995:**18**: 555–586.

144. Bogen JE. On the neurophysiology of consciousness: Part II. Constraining the semantic problem. *Conscious Cogn* 1995:**4**: 137–158.

145. Bogen JE. On the neurophysiology of consciousness: I. An overview. *Conscious Cogn* 1995:**4**: 52–62.

146. Bogen JE. Some neurophysiologic aspects of consciousness. *Semin Neurol* 1997:**17**: 95–103.

147. Savage LM, Castillo R, Langlais PJ. Effects of lesions of thalamic intralaminar and midline nuclei and internal medullary lamina on spatial memory and object discrimination. *Behav Neurosci* 1998:**112**: 1339–1352.

148. Jones EG. A new view of specific and nonspecific thalamocortical connections. *Adv Neurol* 1998:**77**: 49–71.

149. Jones EG. Viewpoint: the core and matrix of thalamic organization. *Neuroscience* 1998:**85**: 331–345.

150. Jeljeli M, Strazielle C, Caston J, Lalonde R. Effects of centrolateral or medial thalamic lesions on motor coordination and spatial orientation in rats. *Neurosci Res* 2000:**38**: 155–164.

151. Llinas RR, Leznik E, Urbano FJ. Temporal binding via cortical coincidence detection of specific and nonspecific thalamocortical inputs: a voltage-dependent dye-imaging study in mouse brain slices. *Proc Natl Acad Sci U S A* 2002:**99**: 449–454.

152. Niblock MM, Kinney HC, Luce CJ, Belliveau RA, Filiano JJ. The development of the medullary serotonergic system in the piglet. *Auton Neurosci* 2004:**110**: 65–80.

153. Mitchell AS, Dalrymple-Alford JC. Dissociable memory effects after medial thalamus lesions in the rat. *Eur J Neurosci* 2005:**22**: 973–985.

154. McCormick DA, Bal T. Sensory gating mechanisms of the thalamus. *Curr Opin Neurobiol* 1994:**4**: 550–556.

155. Newman J, Grace AA. Binding across time: the selective gating of frontal and hippocampal systems modulating working memory and attentional states. *Conscious Cogn* 1999:**8**: 196–212.

156. Schiff ND, Plum F. The role of arousal and "gating" systems in the neurology of impaired consciousness. *J Clin Neurophysiol* 2000:**17**: 438–452.

157. Villablanca J, Salinas-Zeballos ME. Sleep-wakefulness, EEG and behavioral studies of chronic cats without the thalamus: the 'athalamic' cat. *Arch Ital Biol* 1972:**110**: 383–411.

158. Villablanca JR. Counterpointing the functional role of the forebrain and of the brainstem in the control of the sleep-waking system. *J Sleep Res* 2004:**13**: 179–208.

159. Machado-Curbelo C. [Do we defend a brain oriented view of death?]. *Rev Neurol* 2002:**35**: 387–396.

160. Aston-Jones G. Brain structures and receptors involved in alertness. *Sleep Med* 2005:**6 Suppl 1**: S3–S7.
161. Reinscheid RK, Xu YL. Neuropeptide S and its receptor: a newly deorphanized G protein-coupled receptor system. *Neuroscientist* 2005:**11**: 532–538.
162. Nothacker HP, Clark S. From heart to mind. The urotensin II system and its evolving neurophysiological role. *FEBS J* 2005:**272**: 5694–5702.
163. de Lecea L, Sutcliffe JG. The hypocretins and sleep. *FEBS J* 2005:**272**: 5675–5688.
164. Martynska L, Wolinska-Witort E, Chmielowska M, Bik W, Baranowska B. The physiological role of orexins. *Neuro Endocrinol Lett* 2005:**26**: 289–292.
165. Harris GC, Wimmer M, Aston-Jones G. A role for lateral hypothalamic orexin neurons in reward seeking. *Nature* 2005:**437**: 556–559.
166. Achard S, Salvador R, Whitcher B, Suckling J, Bullmore E. A resilient, low-frequency, small-world human brain functional network with highly connected association cortical hubs. *J Neurosci* 2006:**26**: 63–72.
167. Mottaghy FM, Willmes K, Horwitz B, Muller HW, Krause BJ, Sturm W. Systems level modeling of a neuronal network subserving intrinsic alertness. *Neuroimage* 2006:**29**: 225–233.
168. Mountcastle VB. An organization principle for cerebral function: the unit module and the distributed system. In: Edelman GM, Mountcastle V (eds). *The mindful brain*. Cambridge: The MIT Press, 1978; 7–50.
169. Isseroff A, Schwartz ML, Dekker JJ, Goldman-Rakic PS. Columnar organization of callosal and associational projections from rat frontal cortex. *Brain Res* 1984:**293**: 213–223.
170. Albright TD, Desimone R, Gross CG. Columnar organization of directionally selective cells in visual area MT of the macaque. *J Neurophysiol* 1984:**51**: 16–31.
171. Llinas RR, Pare D. Of dreaming and wakefulness. *Neuroscience* 1991:**44**: 521–535.
172. Llinas R, Ribary U. Coherent 40-Hz oscillation characterizes dream state in humans. *Proc Natl Acad Sci U S A* 1993:**90**: 2078–2081.
173. Pare D, Llinas R. Conscious and pre-conscious processes as seen from the standpoint of sleep-waking cycle neurophysiology. *Neuropsychologia* 1995:**33**: 1155–1168.
174. Llinas R, Lang EJ, Welsh JP, Makarenko VI. A New Approach to the Analysis of Multidimensional Neuronal Activity: Markov Random Fields. *Neural Netw* 1997:**10**: 785–789.
175. Hund M, Rezai AR, Kronberg E *et al*. Magnetoencephalographic mapping: basic of a new functional risk profile in the selection of patients with cortical brain lesions. *Neurosurgery* 1997:**40**: 936–942.
176. Llinas R, Ribary U, Contreras D, Pedroarena C. The neuronal basis for consciousness. *Philos Trans R Soc Lond B Biol Sci* 1998:**353**: 1841–1849.
177. Llinas RR, Ribary U, Jeanmonod D, Kronberg E, Mitra PP. Thalamocortical dysrhythmia: A neurological and neuropsychiatric syndrome characterized by magnetoencephalography. *Proc Natl Acad Sci U S A* 1999:**96**: 15222–15227.
178. Ribary U, Cappell J, Mogilner A, Hund-Georgiadis M, Kronberg E, Llinas R. Functional imaging of plastic changes in the human brain. *Adv Neurol* 1999:**81**: 49–56.
179. Llinas RR. Mapping brain terrain. *Neurobiol Dis* 2000:**7**: 499–500.
180. Contreras D, Llinas R. Voltage-sensitive dye imaging of neocortical spatiotemporal dynamics to afferent activation frequency. *J Neurosci* 2001:**21**: 9403–9413.

181. Llinas R, Ribary U. Consciousness and the brain. The thalamocortical dialogue in health and disease. *Ann N Y Acad Sci* 2001:**929**: 166–175.
182. Llinas RR. [Thalamo-cortical dysrhythmia syndrome: neuropsychiatric features]. *An R Acad Nac Med (Madr)* 2003:**120**: 267–290.
183. Jeanmonod D, Schulman J, Ramirez R *et al.* Neuropsychiatric thalamocortical dysrhythmia: surgical implications. *Neurosurg Clin N Am* 2003:**14**: 251–265.
184. Llinas R, Urbano FJ, Leznik E, Ramirez RR, van Marle HJ. Rhythmic and dysrhythmic thalamocortical dynamics: GABA systems and the edge effect. *Trends Neurosci* 2005:**28**: 325–333.
185. Rhodes PA, Llinas R. A model of thalamocortical relay cells. *J Physiol* 2005:**565**: 765–781.
186. Dehaene S, Changeux JP. Ongoing spontaneous activity controls access to consciousness: a neuronal model for inattentional blindness. *PLoS Biol* 2005:**3**: e141.
187. Chabriat H, Pappata S, Levasseur M, Fiorelli M, Tran DS, Baron JC. Cortical metabolism in posterolateral thalamic stroke: PET study. *Acta Neurol Scand* 1992:**86**: 285–290.
188. Feinberg TE. The irreducible perspectives of consciousness. *Semin Neurol* 1997:**17**: 85–93.
189. Mizumori SJ, Yeshenko O, Gill KM, Davis DM. Parallel processing across neural systems: implications for a multiple memory system hypothesis. *Neurobiol Learn Mem* 2004:**82**: 278–298.
190. Wassle H. Parallel processing in the mammalian retina. *Nat Rev Neurosci* 2004:**5**: 747–757.
191. Rousselet GA, Thorpe SJ, Fabre-Thorpe M. How parallel is visual processing in the ventral pathway? *Trends Cogn Sci* 2004:**8**: 363–370.
192. Inui K, Okamoto H, Miki K, Gunji A, Kakigi R. Serial and parallel processing in the human auditory cortex: a magnetoencephalographic study. *Cereb Cortex* 2006:**16**: 18–30.
193. Shewmon DA. "Brain Death": A valid theme with invalid variations, blurred by semantic ambiguity. In: Angstwurm H, Carrasco de Paula I (eds). *Working Group on The Determination of Brain Death and its Relationship to Human Death*. Vatican City: Pontificia Academia Scientiarum, 1992; 23–51.
194. Shewmon DA. The metaphysics of brain death, persistent vegetative state and dementia. *Thomist* 1985:**49**: 24–80.
195. Shewmon DA, Holmes GL, Byrne PA. Consciousness in congenitally decorticate children: developmental vegetative state as self-fulfilling prophecy. *Dev Med Child Neurol* 1999:**41**: 364–374.
196. Shewmon DA. "Brainstem death," "brain death" and death: a critical re-evaluation of the purported equivalence. *Issues Law Med* 1998:**14**: 125–145.
197. Hassler R, Ore GD, Bricolo A, Dieckmann G, Dolce G. EEG and clinical arousal induced by bilateral long-term stimulation of pallidal systems in traumatic vigil coma. *Electroencephalogr Clin Neurophysiol* 1969:**27**: 689–690.
198. Hassler R, Ore GD, Dieckmann G, Bricolo A, Dolce G. Behavioural and EEG arousal induced by stimulation of unspecific projection systems in a patient with post-traumatic apallic syndrome. *Electroencephalogr Clin Neurophysiol* 1969:**27**: 306–310.
199. Sturm V, Kuhner A, Schmitt HP, Assmus H, Stock G. Chronic electrical stimulation of the thalamic unspecific activating system in a patient with coma due to midbrain and upper brain stem infarction. *Acta Neurochir (Wien)* 1979:**47**: 235–244.

200. Cohadon F, Richer E. [Deep cerebral stimulation in patients with post-traumatic veg-etative state. 25 cases]. *Neurochirurgie* 1993:**39**: 281–292.
201. Deliac P, Richer E, Berthomieu J, Paty J, Cohadon F, Bensch C. [Electrophysiological development under thalamic stimulation of post-traumatic persistent vegetative states. Apropos of 25 cases]. *Neurochirurgie* 1993:**39**: 293–303.
202. Yamamoto T, Katayama Y, Oshima H, Fukaya C, Kawamata T, Tsubokawa T. Deep brain stimulation therapy for a persistent vegetative state. *Acta Neurochir Suppl* 2002:**79**: 79–82.
203. Pincus DI. Global neurodynamics and deep brain stimulation: appreciating the per-spectives of place and process. *Adv Exp Med Biol.* 2004;550:239–53.
204. Yamamoto T, Katayama Y, Kobayashi K, Kasai M, Oshima H, Fukaya C. DBS therapy for a persistent vegetative state: ten years follow-up results. *Acta Neurochir Suppl* 2003:**87**: 15–18.
205. Yamamoto T, Katayama Y. Deep brain stimulation therapy for the vegetative state. *Neuropsychol Rehabil* 2005:**15**: 406–413.
206. Yamamoto T, Kobayashi K, Kasai M, Oshima H, Fukaya C, Katayama Y. DBS therapy for the vegetative state and minimally conscious state. *Acta Neurochir Suppl* 2005:**93**: 101–104.
207. Katayama Y, Tsubokawa T, Yamamoto T, Hirayama T, Miyazaki S, Koyama S. Char-acterization and modification of brain activity with deep brain stimulation in patients in a persistent vegetative state: pain-related late positive component of cerebral evoked potential. *Pacing Clin Electrophysiol* 1991:**14**: 116–121.
208. Okun MS, Raju DV, Walter BL *et al.* Pseudobulbar crying induced by stimulation in the region of the subthalamic nucleus. *J Neurol Neurosurg Psychiatry* 2004:**75**: 921–923.
209. Giacino J, Whyte J. The vegetative and minimally conscious states: current knowledge and remaining questions. *J Head Trauma Rehabil* 2005:**20**: 30–50.
210. Duchowny MS. Hemispherectomy and epileptic encephalopathy. *Epilepsy Curr* 2004:**4**: 233–235.
211. Hausmann M, Corballis MC, Fabri M, Paggi A, Lewald J. Sound lateralization in subjects with callosotomy, callosal agenesis, or hemispherectomy. *Brain Res Cogn Brain Res* 2005:**25**: 537–546.
212. Korkman M, Granstrom ML, Kantola-Sorsa E *et al.* Two-year follow-up of intelligence after pediatric epilepsy surgery. *Pediatr Neurol* 2005:**33**: 173–178.
213. de Almeida AN, Marino R, Jr. The early years of hemispherectomy. *Pediatr Neurosurg* 2005:**41**: 137–140.
214. de Almeida AN, Marino R Jr, Aguiar PH, Jacobsen Teixeira M. Hemispherectomy: a schematic review of the current techniques. *Neurosurg Rev* 2006;29(2):97–102.
215. Austin GM, Grant FC. Physiologic observations following total hemispherectomy in man. Surgery 38, 239–258. 1958.
216. O'Brien DF, Basu S, Williams DH, May PL. Anatomical hemispherectomy for in-tractable seizures: excellent seizure control, low morbidity and no superficial cerebral haemosiderosis. *Childs Nerv Syst* 2006: 1–10.
217. Schiff ND, Rodriguez-Moreno D, Kamal A *et al.* fMRI reveals large-scale network activation in minimally conscious patients. *Neurology* 2005:**64**: 514–523.
218. Steriade M, Amzica F, Contreras D. Synchronization of fast (30–40 Hz) spontaneous cortical rhythms during brain activation. *J Neurosci* 1996:**16**: 392–417.
219. Llinas RR, Ribary U. Temporal conjunction in thalamocortical transactions. *Adv Neu-rol* 1998:**77**: 95–102.

220. Shewmon DA. The probability of inevitability: the inherent impossibility of validating criteria for brain death or 'irreversibility' through clinical studies. *Stat Med* 1987:**6**: 535–553.

221. Korein J. Ontogenesis of the fetal nervous system: the onset of brain life. *Transplant Proc* 1990:**22**: 982–983.

222. Korein J. The diagnosis of brain death. *Semin Neurol* 1984:**4**: 52–72.

223. Korein J. The problem of brain death: development and history. *Ann N Y Acad Sci* 1978:**315**: 19–38.

224. Korein J. Brain death: interrelated medical and social issues. Terminology, definitions, and usage. *Ann N Y Acad Sci* 1978:**315**: 6–18.

225. Korein J. Brain death: interrelated medical and social issues. Preface. *Ann N Y Acad Sci* 1978:**315**: 1–5.

226. Bernat JL. Ethical and legal aspects of the emergency management of brain death and organ retrieval. *Emerg Med Clin North Am* 1987:**5**: 661–676.

227. Bernat JL. Ethical issues in brain death and multiorgan transplantation. *Neurol Clin* 1989:**7**: 715–728.

228. Bernat JL, Grabowski EW. Suspending do-not-resuscitate orders during anesthesia and surgery. *Surg Neurol* 1993:**40**: 7–9.

229. Gert B, Bernat JL, Mogielnicki RP. The distinction between active and passive euthanasia. *Arch Intern Med* 1995:**155**: 1329.

230. Bernat JL. Physician-assisted suicide should not be legalized. *Arch Neurol* 1996:**53**: 1183–1184.

231. Bernat JL. The problem of physician-assisted suicide. *Semin Neurol* 1997:**17**: 271–279.

232. Bernat JL. Areas of consensus in withdrawing life-sustaining treatment in the neurointensive care unit. *Neurology* 1999:**52**: 1538–1539.

233. Wijdicks EF, Bernat JL. Chronic "brain death": meta-analysis and conceptual consequences. *Neurology* 1999:**53**: 1369–1370.

234. Bernat JL. The biophilosophical basis of whole-brain death. *Soc Philos Policy* 2002:**19**: 324–342.

235. Bernat JL. T Are organ donors after cardiac death really dead? J Clin Ethics 2006;17(2):122–32.

236. Bernat JL, D'Alessandro AM, Port FK *et al.* Report of a national conference on donation after cardiac death. *Am J Transplant* 2006:**6**: 281–291.

237. A definition of irreversible coma. Report of the Ad Hoc Committee of the Harvard Medical School to Examine the Definition of Brain Death. *JAMA* 1968:**205**: 337–340.

238. Landmark article Aug 5, 1968: A definition of irreversible coma. Report of the Ad Hoc Committee of the Harvard Medical School to examine the definition of brain death. *JAMA* 1984:**252**: 677–679.

239. Uematsu S, Walker AE. A method for recording the pulsation of the midline echo in clinical brain death. *Johns Hopkins Med J* 1974:**135**: 383–390.

240. Moseley JI, Molinari GF, Walker AE. Respirator brain. Report of a survey and review of current concepts. *Arch Pathol Lab Med* 1976:**100**: 61–64.

241. Walker AE. Pathology of brain death. *Ann N Y Acad Sci* 1978:**315**: 272–280.

242. Uematsu S, Smith TD, Walker AE. Pulsatile cerebral echo in diagnosis of brain death. *J Neurosurg* 1978:**48**: 866–875.

243. Walker AE. Current concepts of brain death. *J Neurosurg Nurs* 1983:**15**: 261–264.

244. Walker AE. Brain death—an American viewpoint. *Neurosurg Rev* 1989:**12 Suppl 1**: 259–264.

245. Darby J, Yonas H, Brenner RP. Brainstem death with persistent EEG activity: evaluation by xenon-enhanced computed tomography. *Crit Care Med* 1987:**15**: 519–521.
246. Zwarts MJ, Kornips FH, Vogels OM. Clinical brainstem death with preserved electroencephalographic activity and visual evoked response. *Arch Neurol* 2001:**58**: 1010.
247. Kozlov IA, Sazontseva IE, Moisiuk I, Ermolenko AE, Il'nitskii VV. [Neuroendocrine disorders in brain-dead donors at the time of multiorgan harvesting]. *Anesteziol Reanimatol* 1992: 52–56.
248. Rap ZM. Influence of damage of hypothalamo-hypophysial neurosecretory system and adrenalectomy on the blood-brain barrier alteration during intracranial hypertension. *Acta Neuropathol Suppl (Berl)* 1981:**7**: 10–12.
249. Outwater KM, Rockoff MA. Diabetes insipidus accompanying brain death in children. *Neurology* 1984:**34**: 1243–1246.
250. Hagl C, Szabo G, Sebening C *et al.* Is the brain death related endocrine dysfunction an indication for hormonal substitution therapy in the early period ? *Eur J Med Res* 1997:**2**: 437–440.
251. Szabo G, Hackert T, Sebening C, Melnitchuk S, Vahl CF, Hagl S. Role of neural and humoral factors in hyperdynamic reaction and cardiac dysfunction following brain death. *J Heart Lung Transplant* 2000:**19**: 683–693.
252. Lugo N, Silver P, Nimkoff L, Caronia C, Sagy M. Diagnosis and management algorithm of acute onset of central diabetes insipidus in critically ill children. *J Pediatr Endocrinol Metab* 1997:**10**: 633–639.
253. Pallis C. Television and brain death. *Br Med J* 1980:**281**: 1064.
254. Pallis C. Diagnosis of brain death. *Br Med J* 1980:**281**: 1491–1492.
255. Pallis C. ABC of brain stem death. Pitfalls and safeguards. *Br Med J (Clin Res Ed)* 1982:**285**: 1720–1722.
256. Pallis C. ABC of brain stem death. Diagnosis of brain stem death–II. *Br Med J (Clin Res Ed)* 1982:**285**: 1641–1644.
257. Pallis C. ABC of brain stem death. Diagnosis of brain stem death–I. *Br Med J (Clin Res Ed)* 1982:**285**: 1558–1560.
258. Pallis C. ABC of brain stem death. From brain death to brain stem death. *Br Med J (Clin Res Ed)* 1982:**285**: 1487–1490.
259. Pallis C. ABC of brain stem death. Reappraising death. *Br Med J (Clin Res Ed)* 1982:**285**: 1409–1412.
260. Pallis C. ABC of brain stem death. The arguments about the EEG. *Br Med J (Clin Res Ed)* 1983:**286**: 284–287.
261. Pallis C. ABC of brain stem death. Prognostic significance of a dead brain stem. *Br Med J (Clin Res Ed)* 1983:**286**: 123–124.
262. Pallis C. ABC of brain stem death. The declaration of death. *Br Med J (Clin Res Ed)* 1983:**286**: 39.
263. Pallis CA, Prior PF. Guidelines for the determination of death. *Neurology* 1983:**33**: 251–252.
264. Pallis C. Diabetes insipidus with brain death. *Neurology* 1985:**35**: 1086–1087.
265. Pallis C. Brainstem death: the evolution of a concept. *Semin Thorac Cardiovasc Surg* 1990:**2**: 135–152.
266. Pernick MS. Back from the grave: Recurring controversies over defining and diagnosing death in history. In: RM Zaner (ed). *Death: Beyond the Whole Brain Criteria.* New York: Kluwer Academics, 1988; 17–74.
267. Rodin E, Tahir S, Austin D, Andaya L. Brainstem death. *Clin Electroencephalogr* 1985:**16**: 63–71.

268. Ferbert A, Buchner H, Ringelstein EB, Hacke W. Isolated brain-stem death. Case report with demonstration of preserved visual evoked potentials (VEPs). *Electroencephalogr Clin Neurophysiol* 1986:**65**: 157–160.

269. Alarcon DG. [Clinical death. Legal death; somatic death; complete death]. *Rev Invest Clin* 1972:**24**: 213–220.

270. Brain death with prolonged somatic survival. *N Engl J Med* 1982:**306**: 1361–1363.

271. Takatori T. [Brain death and somatic death]. *Hokkaido Igaku Zasshi* 1987:**62**: 682–686.

272. Powner DJ, Bernstein IM. Extended somatic support for pregnant women after brain death. *Crit Care Med* 2003:**31**: 1241–1249.

273. Mallampalli A, Powner DJ, Gardner MO. Cardiopulmonary resuscitation and somatic support of the pregnant patient. *Crit Care Clin* 2004:**20**: 747–61.

274. Farragher RA, Laffey JG. Maternal brain death and somatic support. *Neurocrit Care* 2005:**3**: 99–106.

275. Field DR, Gates EA, Creasy RK, Jonsen AR, Laros RK, Jr. Maternal brain death during pregnancy. Medical and ethical issues. *JAMA* 1988:**260**: 816–822.

276. Lane A, Westbrook A, Grady D *et al.* Maternal brain death: medical, ethical and legal issues. *Intensive Care Med* 2004:**30**: 1484–1486.

277. Farragher R, Marsh B, Laffey JG. Maternal brain death—an Irish perspective. *Ir J Med Sci* 2005:**174**: 55–59.

278. Fisher J. Re-examining death: against a higher brain criterion. *J Med Ethics* 1999:**25**: 473–476.

279. DeGrazia D. Persons, organisms, and death: a philosophical critique of the higher-brain approach. *South J Philos* 1999:**37**: 419–440.

280. DuBois JM, Schmidt T. Does the public support organ donation using higher brain-death criteria? *J Clin Ethics* 2003:**14**: 26–36.

281. Lowinger P, Fletcher J, Veatch RM. Brain death case. *Hastings Cent Rep* 1973:**3**: 12.

282. Veatch RM. Life and Death Decision Making, by Baruch A. Brody. *Bioethics* 1989:**3**: 147–151.

283. Giacino JT, Ashwal S, Childs N *et al.* The minimally conscious state: definition and diagnostic criteria. *Neurology* 2002:**58**: 349–353.

284. Puccetti R. Does anyone survive neocortical death? In: Zaner RM (ed). *Death: Beyond Whole Brain Criteria*. Boston: Kluwer Academics, 1988; 75–90.

285. Veatch RM. *Death, dying, and the biological revolution. Our last responsibility*. New Haven: Yale University Press,1989.

286. Lizza JP. Persons and death: what's metaphysically wrong with our current statutory definition of death? *J Med Philos* 1993:**18**: 351–374.

287. Lizza JP. The conceptual basis for brain death revisited: loss of organic integration or loss of consciousness? *Adv Exp Med Biol* 2004:**550**: 51–59.

288. Lizza JP. Potentiality, irreversibility, and death. *J Med Philos* 2005:**30**: 45–64.

289. Machado C, García-Tigera J, García O, García-Pumariega J, and Román J. Muerte Encefálica. Criterios diagnósticos. *Revista Cubana de Medicina* 30(3), 181–206. 1991.

290. Machado-Curbelo C. [A new formulation of death: definition, criteria and diagnostic tests]. *Rev Neurol* 1998:**26**: 1040–1047.

291. Machado C. The minimally conscious state: definition and diagnostic criteria. *Neurology* 2002:**59**: 1473–1474.

292. Machado C. A definition of human death should not be related to organ transplants. *J Med Ethics* 2003:**29**: 201–202.

293. Machado C, Sherman DL. *Brain Death and Disroders of Consciousness*, Vol 50. New York: Kluwer Academics/Plenum Publishers, 2004.
294. Machado C. Can vegetative state patients retain cortical processing? *Clin Neurophysiol* 2005:**116**: 2253–2254.
295. Machado C. Cerebral processing in the minimally conscious state. *Neurology* 2005:**65**: 973–974.
296. Tsubokawa T, Yamamoto T, Katayama Y, Hirayama T, Maejima S, Moriya T. Deep-brain stimulation in a persistent vegetative state: follow-up results and criteria for selection of candidates. *Brain Inj* 1990:**4**: 315–327.
297. Bernat JL. The boundaries of the persistent vegetative state. *J Clin Ethics* 1992:**3**: 176–180.
298. Kinney HC, Korein J, Panigrahy A, Dikkes P, Goode R. Neuropathological findings in the brain of Karen Ann Quinlan. The role of the thalamus in the persistent vegetative state. *N Engl J Med* 1994:**330**: 1469–1475.
299. Adams JH, Graham DI, Jennett B. The neuropathology of the vegetative state after an acute brain insult. *Brain* 2000:**123 (Pt 7)**: 1327–1338.
300. Bernat JL. Questions remaining about the minimally conscious state. *Neurology* 2002:**58**: 337–338.
301. Giacino JT. The vegetative and minimally conscious states: consensus-based criteria for establishing diagnosis and prognosis. *NeuroRehabilitation* 2004:**19**: 293–298.
302. Giacino JT, Kalmar K, Whyte J. The JFK Coma Recovery Scale-Revised: measurement characteristics and diagnostic utility. *Arch Phys Med Rehabil* 2004:**85**: 2020–2029.
303. Laureys S. The neural correlate of (un)awareness: lessons from the vegetative state. *Trends Cogn Sci* 2005:**9**: 556–559.
304. Machado C, Wagner A, Coutin P, Diaz G, Cantón M, Hernández O, Roman JM, and Miranda J. Potenciales Evocados Somatosensoriales de corta latencia. II- Tiempo de Conducción Central. Revista del Hospital Psiquiátrico de La Habana, 211–221. 1988.
305. Jennett B, Adams JH, Murray LS, Graham DI. Neuropathology in vegetative and severely disabled patients after head injury. *Neurology* 2001:**56**: 486–490.
306. Laureys S, Antoine S, Boly M *et al*. Brain function in the vegetative state. *Acta Neurol Belg* 2002:**102**: 177–185.
307. Jennett B. The vegetative state. *J Neurol Neurosurg Psychiatry* 2002:**73**: 355–357.
308. Schiff ND, Ribary U, Moreno DR *et al*. Residual cerebral activity and behavioural fragments can remain in the persistently vegetative brain. *Brain* 2002:**125**: 1210–1234.
309. Graham DI, Adams JH, Murray LS, Jennett B. Neuropathology of the vegetative state after head injury. *Neuropsychol Rehabil* 2005:**15**: 198–213.
310. Steinbock B. Recovery from persistent vegetative state? The case of Carrie Coons. *Hastings Cent Rep* 1989:**19**: 14–15.
311. Childs NL, Mercer WN, Childs HW. Accuracy of diagnosis of persistent vegetative state. *Neurology* 1993:**43**: 1465–1467.
312. Cohen-Almagor R. Some observations on post-coma unawareness patients and on other forms of unconscious patients: policy proposals. *Med Law* 1997:**16**: 451–471.
313. Giacino JT. Disorders of consciousness: differential diagnosis and neuropathologic features. *Semin Neurol* 1997:**17**: 105–111.
314. Birbaumer N, Ghanayim N, Hinterberger T *et al*. A spelling device for the paralysed. *Nature* 1999:**398**: 297–298.
315. Kotchoubey B, Haisst S, Daum I, Schugens M, Birbaumer N. Learning and self-regulation of slow cortical potentials in older adults. *Exp Aging Res* 2000:**26**: 15–35.

316. Kotchoubey B, Kubler A, Strehl U, Flor H, Birbaumer N. Can humans perceive their brain states? *Conscious Cogn* 2002:**11**: 98–113.

317. Kotchoubey B, Lang S. Parallel processing of physical and lexical auditory information in humans. *Neurosci Res* 2003:**45**: 369–374.

318. Giacino JT, Whyte J. Amantadine to improve neurorecovery in traumatic brain injury-associated diffuse axonal injury: a pilot double-blind randomized trial. *J Head Trauma Rehabil* 2003:**18**: 4–5.

319. Whyte J, Katz D, Long D *et al.* Predictors of outcome in prolonged posttraumatic disorders of consciousness and assessment of medication effects: A multicenter study. *Arch Phys Med Rehabil* 2005:**86**: 453–462.

320. Cicerone KD, Dahlberg C, Malec JF *et al.* Evidence-based cognitive rehabilitation: updated review of the literature from 1998 through 2002. *Arch Phys Med Rehabil* 2005:**86**: 1681–1692.

321. Magee WL. Music therapy with patients in low awareness states: approaches to assessment and treatment in multidisciplinary care. *Neuropsychol Rehabil* 2005:**15**: 522–536.

322. Elliott L, Walker L. Rehabilitation interventions for vegetative and minimally conscious patients. *Neuropsychol Rehabil* 2005:**15**: 480–493.

323. Reimer M, LeNavenec CL. Rehabilitation outcome evaluation after very severe brain injury. *Neuropsychol Rehabil* 2005:**15**: 473–479.

324. Andrews K. Rehabilitation practice following profound brain damage. *Neuropsychol Rehabil* 2005:**15**: 461–472.

325. Wilson BA. Part III: Behavioural assessment and rehabilitation techniques. *Neuropsychol Rehabil* 2005:**15**: 428–430.

326. Botkin JR, Post SG. Confusion in the determination of death: distinguishing philosophy from physiology. *Perspect Biol Med* 1992:**36**: 129–138.

327. Veatch RM. Brain death and slippery slopes. *J Clin Ethics* 1993:**4**: 376.

328. Machado C. Una nueva definición de la de muerte según criterios neurológicos. In: Esteban A, Escalante A (eds). *Muerte Encefálica y Donación de Órganos.* Madrid: Comunidad Autónoma de Madrid, 1995; 27–51.

329. Machado C, García OD, Román JM, Parets J. Four years after the "First International Symposium on Brain Death" in Havana: Could a definitive conceptual re-approach be expected? In: Machado C (ed). *Brain Death (Proceedings of the Second International Symposium on Brain Death)*, Vol 9. Amsterdam: Elsevier Science, B. V, 1995; 1–9.

330. Korein J, Maccario M. On the diagnosis of cerebral death—a prospective study. *Electroencephalogr Clin Neurophysiol* 1969:**27**: 700.

331. Korein J. On cerebral, brain, and systemic death. *Curr Conc Cerebrovasc Dis Stroke* 1973:**8**: 9–14.

332. Swedish Committee on Defining Death. The Concept of Death. Summary. Stockholm: Swedish Ministry of Health and Social Affairs,1985.

333. Madanmohan, Udupa K, Bhavanani AB, Shatapathy CC, Sahai A. Modulation of cardiovascular response to exercise by yoga training. *Indian J Physiol Pharmacol* 2004:**48**: 461–465.

334. Mamtani R, Mamtani R. Ayurveda and Yoga in Cardiovascular Diseases. *Cardiol Rev* 2004:**12**: 155–162.

335. Telles S, Joshi M, Dash M, Raghuraj P, Naveen KV, Nagendra HR. An evaluation of the ability to voluntarily reduce the heart rate after a month of yoga practice. *Integr Physiol Behav Sci* 2004:**39**: 119–125.

336. Yayoi O. [Influences of premenstrual syndrome on daily psychological states and salivary cortisol level]. *Shinrigaku Kenkyu* 2005:**76**: 426–435.
337. Hahn S, Janssen OE, Tan S *et al.* Clinical and psychological correlates of quality-of-life in polycystic ovary syndrome. *Eur J Endocrinol* 2005:**153**: 853–860.
338. Paoletti AM, Romagnino S, Contu R *et al.* Observational study on the stability of the psychological status during normal pregnancy and increased blood levels of neuroactive steroids with GABA-A receptor agonist activity. *Psychoneuroendocrinology* 2006;31(4):485–92.
339. Wikler D. Brain death: a durable consensus? *Bioethics* 1993:**7**: 239–246.
340. Truog RD and Flacker JC. Rethinking brain death. *Critical Care Medicine* 20, 1705–1713. 2006.
341. Vaitl D, Birbaumer N, Gruzelier J *et al.* Psychobiology of altered states of consciousness. *Psychol Bull* 2005:**131**: 98–127.
342. Feinberg TE, Keenan JP. Where in the brain is the self? *Conscious Cogn* 2005:**14**: 661–678.
343. Arduini A, Moruzzi G. Olfactory arousal reactions in the "cerveau isole" cat. *Electroencephalography and Clinical Neurophysiology* 1953:**5**: 243–250.
344. Moruzzi G. Reticular influences on the EEG. *Electroencephalography and Clinical Neurophysiology* 1964:**16**: 2–17.

FIGURE 1.3. Peter Safar during the III International Symposium on Coma and Death, Havana, 2000. Father of modern reanimatology.

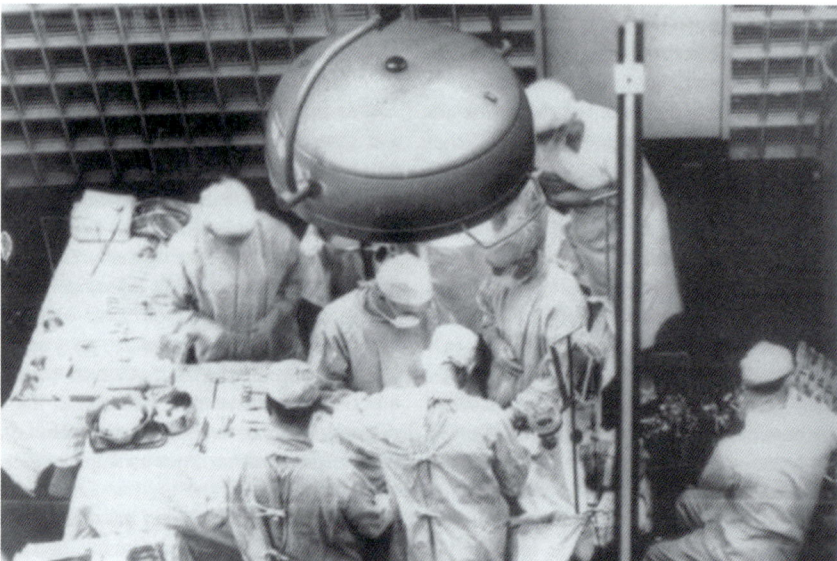

FIGURE 1.10. (A) Joseph E. Murray honored at a special commemorative event celebrating 50 years of transplantation at the U.S. Transplant Games on July 29, 1994. (B) Murray and his team carried out the first successful human organ transplant, taking a kidney from an identical twin on December 23, 1954. Photos credit Eric Miller for the National Kidney Foundation, and the Surgery of the Soul, Science History Publications, Watson Publishing International.

FIGURE 1.11. Christiaan Barnard. Photograph taken during the proceedings of a meeting held on December 28, 1967, at O'Hare Airport, Chicago, Illinois, on the subject of cardiac transplantation. Left to right: David M. Hume, Christiaan Barnard, Andrew G. Morrow, Donald S. Fredrickson. Courtesy of the National Library of Medicine.

FIGURE 2.1. Guy P. J. Alexandre, Belgian, born on July 4, 1934. (Photo provided by Dr. Alexandre.)

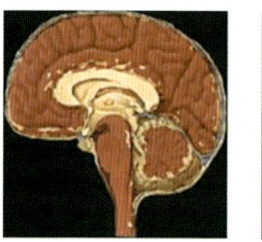 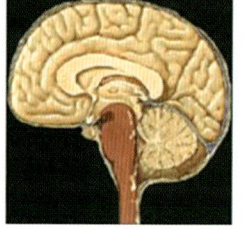 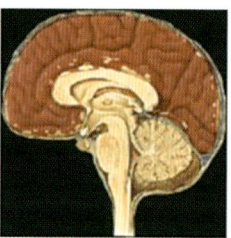

Whole brain   Brainstem death   Higher brain

FIGURE 3.1. Three main brain-oriented formulations of death have been discussed: whole brain, brainstem death, and higher brain standards.

FIGURE 3.3. From right to left: Prof. Fred Plum, Prof. Calixto Machado, Prof. and Mrs. Jerome Posner, and Dr. Yazmina Ferrer (Dr. Machado's wife).

(A)

Horace W. Magoun   Giuseppe Moruzzi

(B)

Cortex

Thalamus

Brainstem

ASCENDING RETICULAR ACTIVATING SYSTEM

FIGURE 3.5. (A) Giuseppe Moruzzi and Horace W. Magoun discovered "the presence in the brainstem of a system of ascending reticular relays, whose direct stimulation activates or desynchronizes the EEG, replacing high-voltage low waves with low voltage fast activity" in 1949. (B) Ascending reticular activating system (ARAS). This system mainly provides arousal, and it is organized by several brainstem and diencephalic neuron populations that project to cortical neurons. Courtesy of the National Library of Medicine.

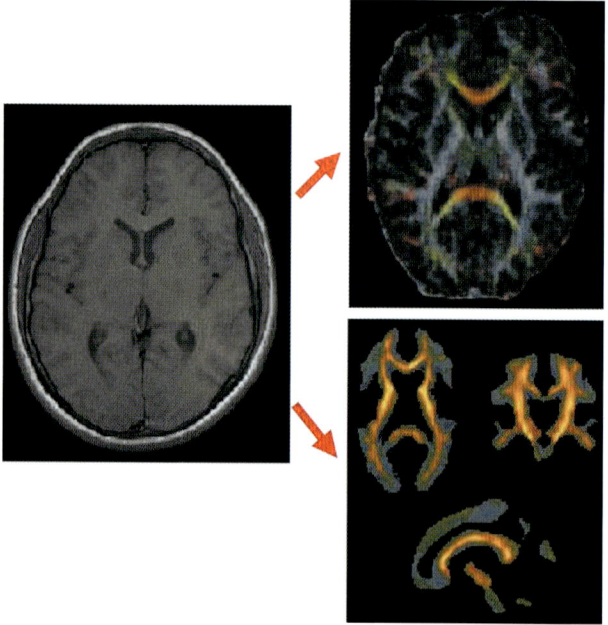

FIGURE 3.6. Cortical columns of the cerebral cortex. Awareness is thought to be dependent on the functional integrity of the cerebral cortex and its subcortical connections.

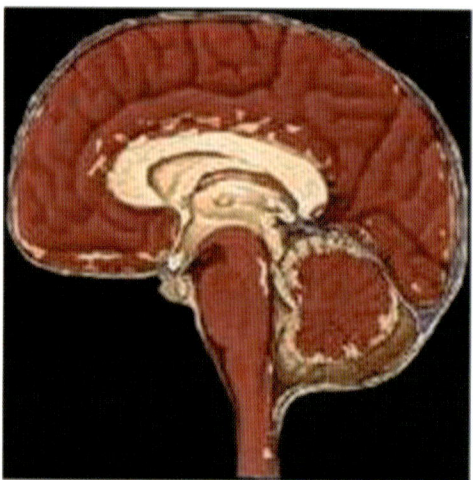

Whole Brain Standard

FIGURE 3.7. Whole brain standard defenders have defined death as "the permanent cessation of the functioning of the organism as a whole."

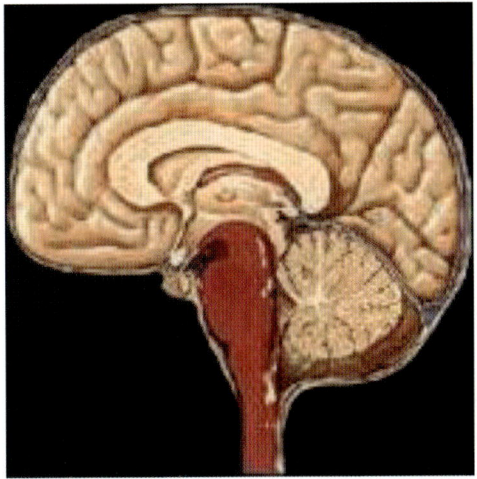

Brainstem Death

FIGURE 3.8. Brainstem death standard defenders have considered that the permanent cessation of the functioning of the brainstem means death.

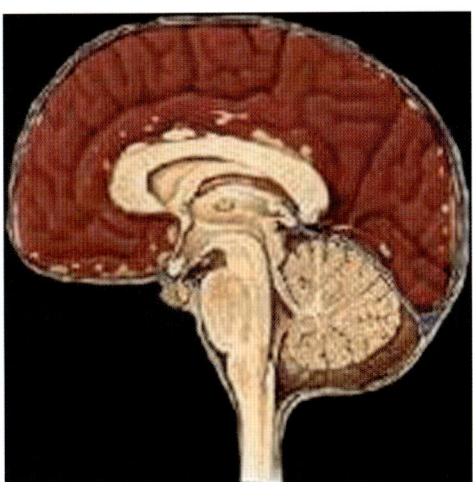

Higher Brain Standard

FIGURE 3.9. Higher brain advocates have proposed defining death as "the loss of that which is significant to the nature of man," which is related only to the cerebral cortex functioning.

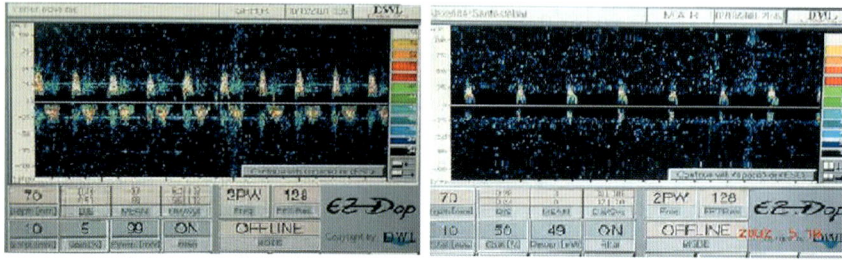

OSCILLATING FLOW　　　　　**SYSTOLIC SPIKES**

FIGURE 5.1. Doppler-sonographic patterns in two brain-dead patients: Left (Oscillating flaw), Right (Systolic spikes). Courtesy of Dr. Anselmo Abdo.

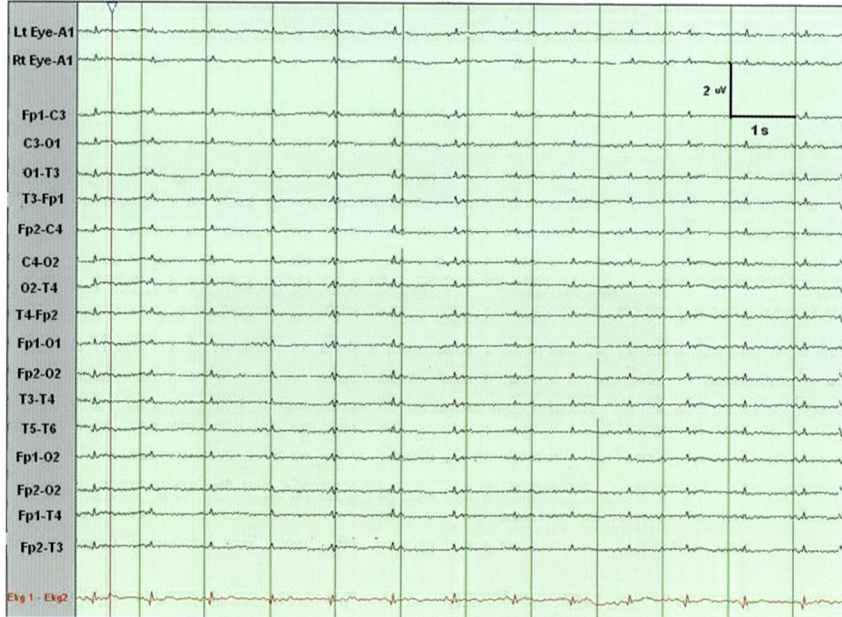

FIGURE 5.2. Electrocerebral silence in a brain-dead patient using digital electroencephalogram (EEG). See electrocardiogram (ECG) contamination in all leads.

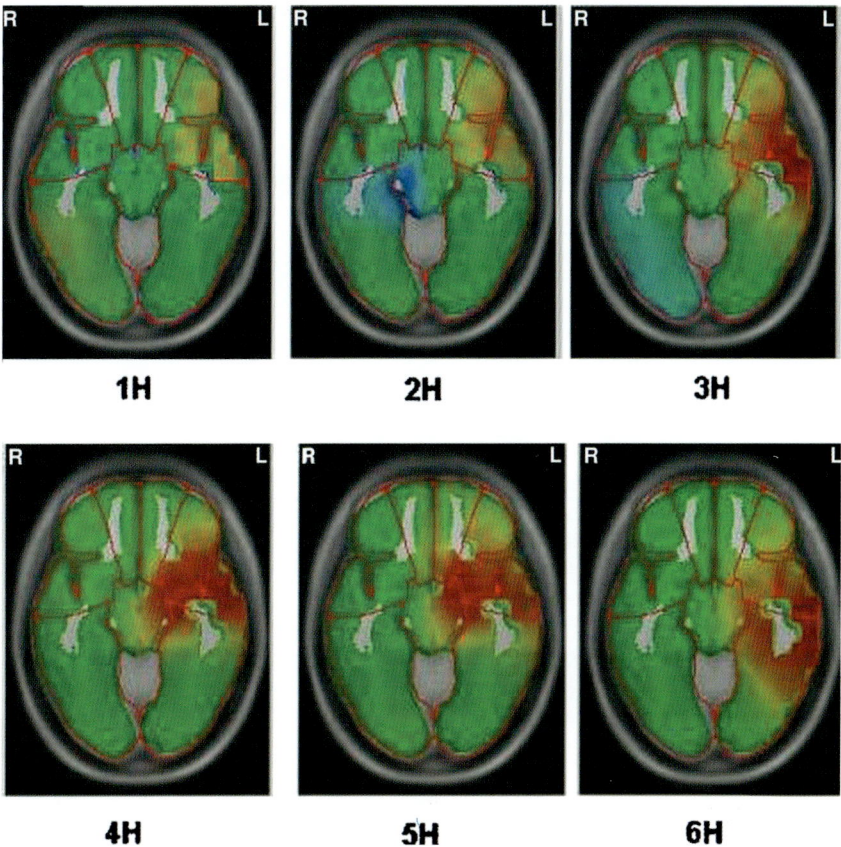

FIGURE 5.3. Continuous monitoring in a comatose patient. At 6 hours, an electrocerebral silence (see Fig. 5.2) is recorded.[256]

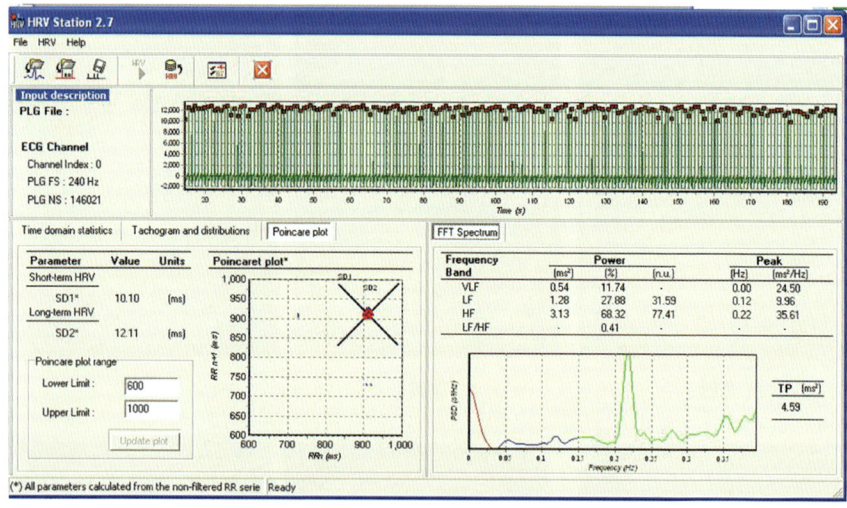

FIGURE 5.11. Heart rate variability in a brain-dead patient. Recorded by the heart rate variability (HRV) station system developed in our laboratory.

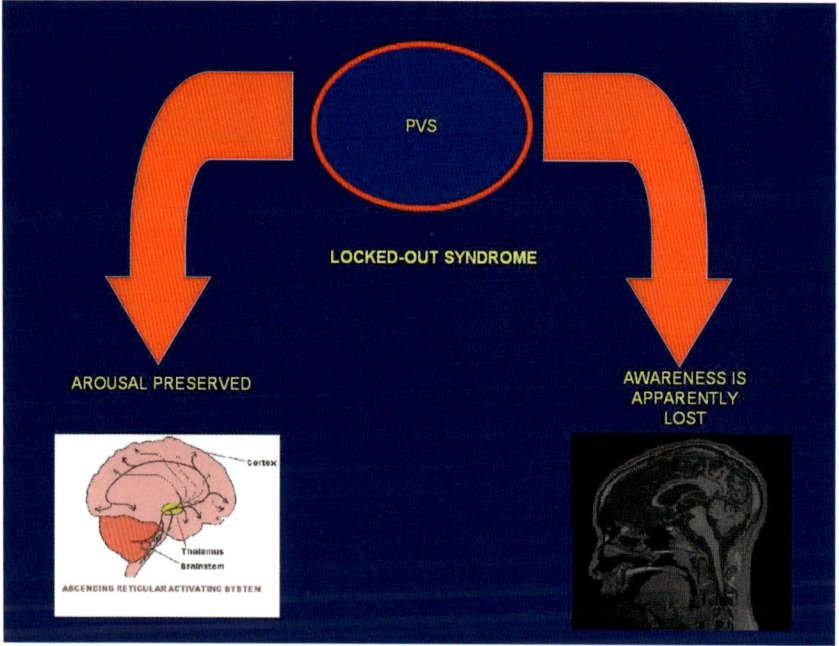

FIGURE 7.1. The main finding in vegetative state (VS) is preservation of wakefulness with apparent loss of awareness.

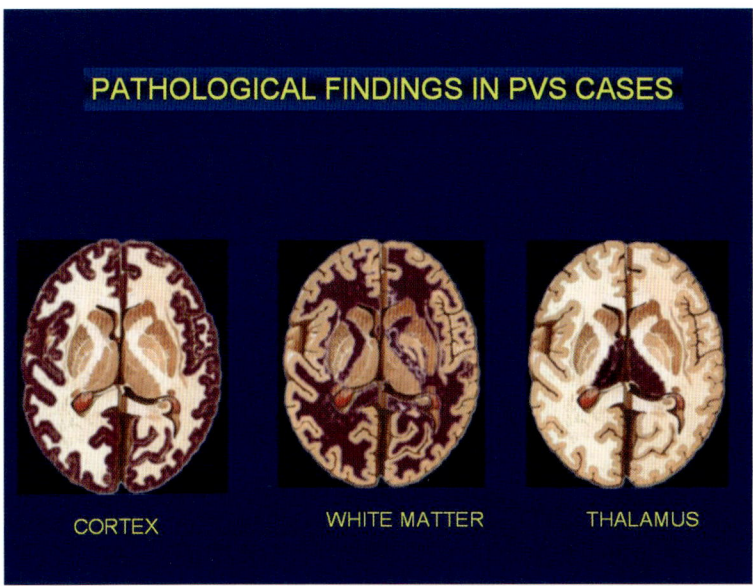

FIGURE 7.2. The VS syndrome is caused by three main patterns: widespread and bilateral lesions of the cerebral cortex, diffuse damage of intra- and subcortical connections in the cerebral hemispheres white matter, and necrosis of the thalamus. (Modified from Kinney et al.[28])

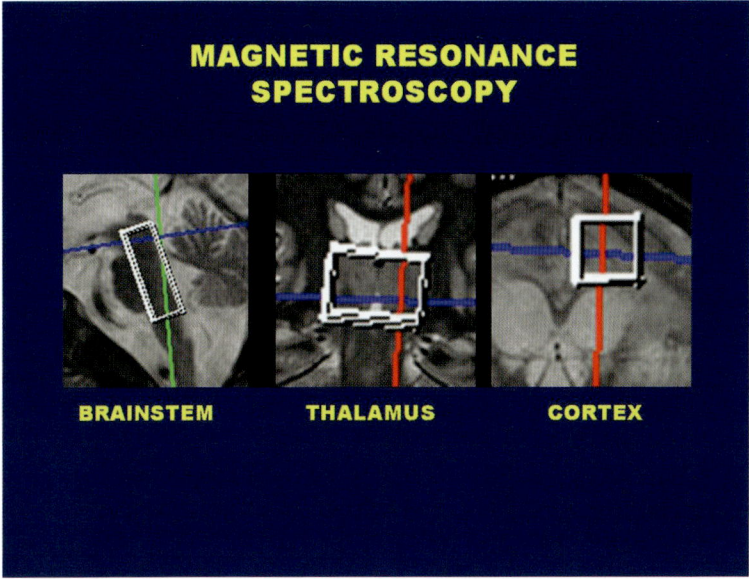

FIGURE 7.3. Regions selected to study magnetic resonance spectroscopy: brainstem, thalamus, and cortex.

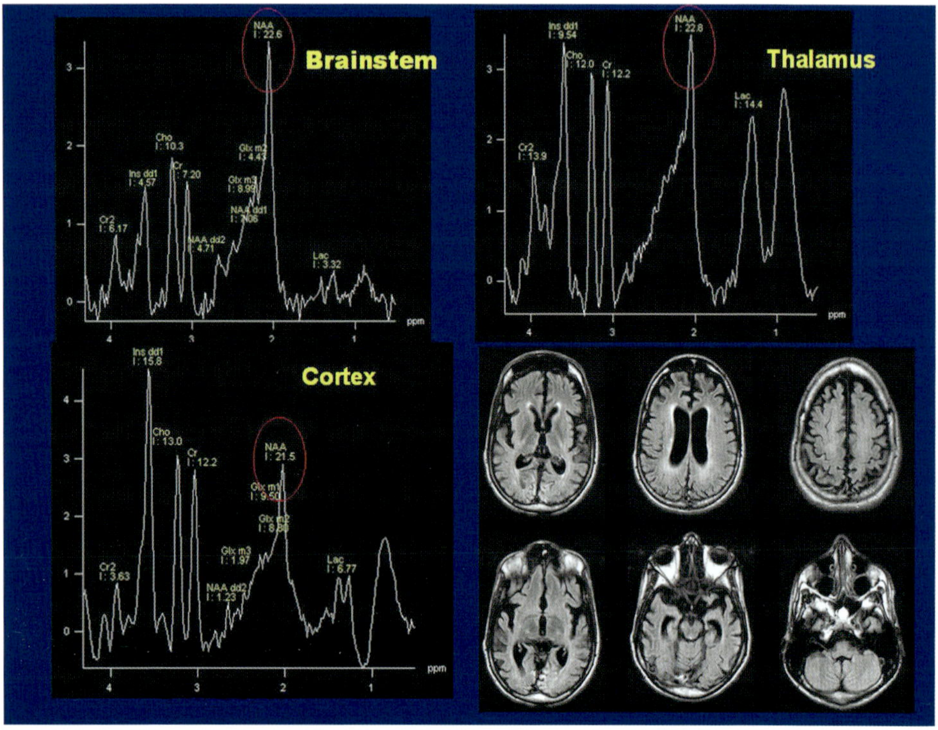

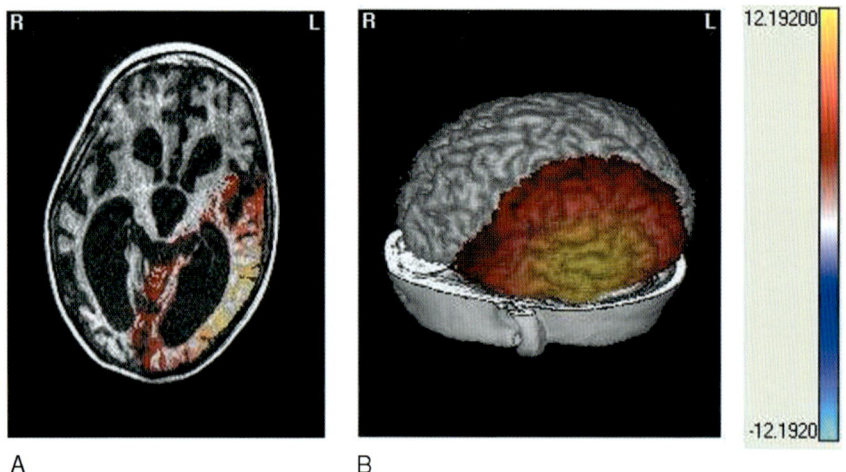

FIGURE 7.4. Magnetic resonance spectroscopy in a PVS case. N-acetylaspartate (NAA) presents a lower concentration in cerebral cortex, compared to brainstem and thalamus, in a PVS patient.

FIGURE 7.5. Quantitative electroencephalographic tomography (QEEGt) is calculated in a PVS patient. QEEGt maps in an axial plane (A) and in a 3D reconstruction. Red and yellow colors show brain regions that might be activated when patient's mother talks to him.

4
Clinical Diagnosis of Brain Death

Most authors affirm that the diagnosis of brain death (BD) is just a clinical assessment, and after the Harvard Committee Report was published,[1,2] most countries and states designed their BD diagnostic criteria.[3−32] Brain death diagnosis should be carried out following a certain set of principles, excluding confusing factors, establishing the cause of the coma, determining irreversibility, and precisely testing brainstem reflexes at all levels of the brainstem.[33]

Nonetheless, most criteria for BD diagnosis do not acknowledge that this is not the only way of diagnosing death. If a concept of death on neurological grounds is accepted, then BD diagnostic criteria can be applied only in patients under life support assistance in intensive care units (ICUs). Does this mean that when a physician diagnoses death in a patient on a regular ward (i.e., the patient is not on life support) applying cardiocirculatory and respiratory diagnostic criteria, or when a forensic specialist diagnoses death in a body under criminal circumstances, are we denying a brain-oriented concept of death?[6,34−37]

Cuba has recently passed a law for the determination and certification of death. The National Commission for the Determination and Certification of Death has accepted a neurological view of death and emphasized that any legislation of death should be completely separated from any norm governing organ transplants. Hence, the commission enumerated three possible situations for diagnosing death[6,36−38]:

1. Outside the intensive care environment (without life support), physicians apply the cardiocirculatory and respiratory criteria.
2. In forensic medicine circumstances, physicians utilize cadaveric signs. (They do not even need a stethoscope.)
3. In the intensive care environment (with life support), when cardiocirculatory or respiratory arrest occurs, physicians utilize the cardiocirculatory and respiratory criteria. When physicians suspect an irreversible loss of brain functions in a heart-beating and ventilatory supported case, BD diagnostic criteria are applied (Fig. 4.1).

This methodology of diagnosing death, based on finding any of the signs of death, is not related to the concept that there are different types of death. The irreversible loss of cardiocirculatory and respiratory functions can only cause death when

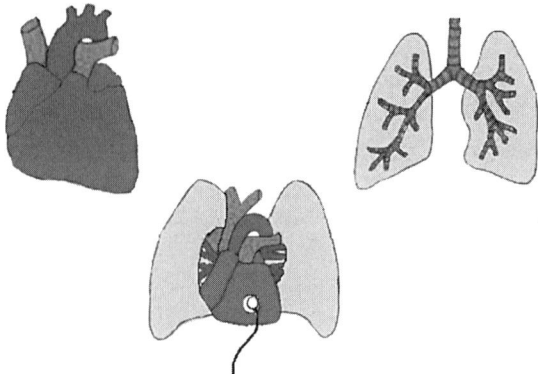

FIGURE 4.1. Sings of irreversible loss of respiratory and cardiocirculatory functions.

ischemia and anoxia are prolonged enough to produce an irreversible destruction of the brain. According to the commission there is only one kind of death, based on the irreversible loss of brain functions.[6,36–38]

Signs of Death

The diagnosis of death was based on the finding of any of these signs of death (Table 4.1):

    I. Irreversible loss of respiratory function
    II. Irreversible loss of cardiocirculatory functions
    III. Algor mortis (postmortem coldness)
    IV. Livor mortis (postmortem lividity)
    V. Rigor mortis (postmortem rigidity)
    VI. Cadaveric spasm
    VII. Loss of muscle contractions
    VIII. Putrefaction
    IX. Irreversible loss of brain functions

Signs I and II correspond to the classic respiratory and cardiocirculatory functions. Signs III to VIII are related to forensic circumstances. These methods are beyond the scope of this chapter. Sign IX corresponds to the BD diagnosis.

Signs I and II: Irreversible Loss of Respiratory and Cardiocirculatory Functions

Cardiac arrest is determined in the field by the absence of a spontaneous pulse palpated at a carotid, and by the auscultation of the main four auscultation areas (aortic, pulmonic, tricuspid, and mitral areas). It also can be determined by

TABLE 4.1. Criteria for brain death diagnosis in adults

Preconditions or prerequisites

1. Coma due to an irreversible acute brain damage of known etiology, affecting both hemispheres and brainstem
2. At the time of beginning neurological examination, the following confounding factors should be excluded:
   a) Unresuscitated shock; systolic arterial pressure should be over 90 mm Hg
   b) Hypothermia (core temperature $<34°C$)
   c) Severe metabolic disorders capable of causing a potentially reversible coma
   d) Peripheral nerve or muscle dysfunction, and neuromuscular blockade
   e) Evidence of drug intoxication or poisoning

Diagnostic criteria

1. Deep unresponsive coma
2. Absent brainstem reflexes
   • Corneal responses
   • Pupillary responses to light with pupils at mid-size or greater
   • Oculocephalic and oculovestibular responses
   • Gag and cough responses
3. Negative atropine test
4. Absent respiratory effort confirmed by the apnea test
5. Periods of observation: 6 hours; in cases of acute hypoxic-ischemic brain injury, clinical evaluation should be delayed for 24 hours subsequent to the cardiorespiratory arrest or an ancillary test could be performed.
6. Confirmatory tests
   • Tests to demonstrate a complete cessation of brain circulation: Transcranial Doppler (TCD) ultrasonography
   • Confirmatory tests to demonstrate loss of bioelectrical activity: electroencephalography (EEG), evoked potential (EP), and electroretinography (ERG)

the presence of ventricular fibrillation or asystole using an electrocardiographic monitor. *Respiratory arrest* is determined in the field by the absence of effective, spontaneous ventilation, assessed by inspection and auscultation (Fig. 4.2).

The diagnosis of death can be completed if a cardiorespiratory arrest of at least 5 minutes is demonstrated, excluding hypothermia. Nonetheless, if an effective mechanical cardiorespiratory reanimation is applied or neuroprotective methods are used (as controlled hypothermia) this 5-minute period could be increasingly prolonged. Several authors have reported patients undergoing accidental hypothermia (immersion/submersion in cold water, snow avalanche, or prolonged exposure to cold surroundings) combined with circulatory arrest or severe circulatory failure who were rewarmed to normothermia by use of extracorporeal circulation, with good outcome in several cases.[39–50] Hence, in all cardiac arrests with exposure to hypothermia, rewarming and reanimation should be attempted.[51]

Sign III: Algor Mortis

Algor mortis (from the Latin: *algor*, meaning coolness; *mortis*, meaning death), also known as body cooling, is the reduction in body temperature following death. This is generally a steady decline until matching ambient temperature, although

THREE POSSIBLE SITUATIONS FOR DIAGNOSING DEATH

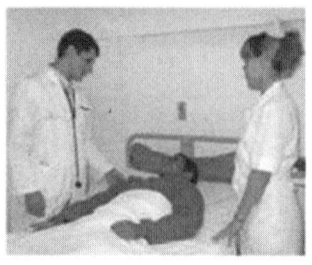 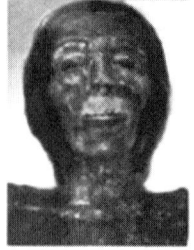 

Hospital Ward	Forensic circumstances	Intensive care unit

FIGURE 4.2. Three possible situations for diagnosing death: without life support, in forensic medicine circumstances, and with life support. When cardiocirculatory or cardiorespiratory arrest occurs, physicians utilize the cardiocirculatory and cardiorespiratory criteria; when physicians suspect an irreversible loss of brain functions in a heart-beating and ventilatory supported patient, BD diagnostic criteria are applied.

external factors can have a significant influence. As decomposition occurs the internal body temperature tends to rise again. This is the most useful single indicator of the time of death during the first 24 hours postmortem. The Glaister equation, for example, is as follows: (98.4°F − rectal temperature)/1.5. The resulting number is the hours elapsed since death.[52–56]

Sign IV: Livor Mortis

Livor mortis (hypostasis, postmortem lividity, postmortem suggillations) is a settling of the blood in the lower (dependent) portion of the body, causing a dark purplish red discoloration of the skin resulting from the gravitational pooling of blood in the veins and capillary beds of the dependent parts of the body following cessation of the circulation; when the heart is no longer agitating the blood, heavy red blood cells sink through the serum by action of gravity. This discoloration does not occur in the areas of the body that are in contact with the ground or another object, as they are compressed. It appears about 1 to 2 hours after death and becomes fixed within 8 to 10 hours, and in a person dying slowly with circulatory failure it may be pronounced very shortly after death. Lividity is present in all bodies, although it may be inconspicuous in some cadavers.[57–65]

Sign V: Rigor Mortis

Rigor mortis, also known as postmortem rigidity, is the muscular stiffening subsequent to death. The onset of rigor mortis varies from 10 minutes to several hours following death and is dependent on atmospheric factors, particularly temperature. Rigor mortis persists for approximately 24 hours or until the actin–myosin links are destroyed by autolysis. Rigor mortis affects the facial musculature first and then spreads to other parts of the body. It is caused by chemical changes in the muscle tissue. Depending on temperature and other conditions, rigor mortis lasts approximately 72 hours. Following death, cisternal and extracellular calcium leak back into the muscle cells' sarcoplasm. This raises the calcium concentration to high levels and causes conformational changes in the troponin-tropomyosin complex associated with actin filaments. The calcium binds to the troponin-tropomyosin complex and causes the exposure of the myosin binding sites on the thin actin filaments. Myosin now interacts with its corresponding site on the actin filaments. When the glycogen store is depleted, the adenosine triphosphate (ATP) concentration falls and the muscle will now become rigid due to the permanent link between actin and myosin. Rigor mortis can be used to help estimate time of death. The onset of rigor mortis may range from 10 minutes to several hours, depending on factors including temperature (rapid cooling of a body can inhibit rigor mortis, but it occurs upon thawing). Maximum stiffness is reached around 12 to 24 hours postmortem. Facial muscles are affected first, with the rigor then spreading to other parts of the body. The joints are stiff for 1 to 3 days, but after this time general tissue decay and leaking of lysosomal intracellular digestive enzymes cause the muscles to relax.[59,66-77]

Sign VI: Cadaveric Spasm

Cadaveric spasm refers to a kind of instant rigor mortis. After a person dies, the muscles of the body stiffen, from using up their oxygen supply and because wastes—carbon dioxide and urea—are no longer carried away from the muscle cells. It also has to do with the depletion of muscle enzymes. So cadaveric spasm is a muscle phenomenon in which some muscles of the body become stiff instantly, rather than in the usual 2 to 8 hours normal rigor takes to develop. The reason for such rapid rigor mortis, usually, is the extreme exertion of the muscles during the act of dying, especially as can happen during a struggle.[67,70,75-77]

Sign VII: Loss of Muscle Contraction

This is a late sign after death characterized by loss of muscle contraction to electric or mechanical stimuli. At the time of death, a few muscle contractions may occur, and the chest may heave as if to breathe.[67,70,77,78]

Sign VIII: Putrefaction

Putrefaction (postmortem decomposition) is the postmortem destruction of the soft tissues of the body by the action of bacteria and enzymes (both bacterial and endogenous), occurring immediately upon death. Before the advent of the stethoscope, this was classified by some authors as the most secure sign of death.[79]

Tissue breakdown resulting from the action of endogenous enzymes alone is known as autolysis. Putrefaction results in the gradual dissolution of the tissues into gases, liquids, and salts. The main changes that can be recognized in the tissues undergoing putrefaction are changes in color, the evolution of gases, and liquefaction. Eventually the skin blisters and fills with fluid or gas, producing a very disagreeable odor. The face darkens and liquids escape the nose and mouth. The tongue swells and the abdomen begins to turn a greenish-yellow color. If the weather is warm and humid, putrefaction may set in within a day, but if the body is left in a very cold area or storage space, putrefaction may be retarded for several months. Environmental temperature has a great influence on the rate of development of putrefaction. Putrefaction is optimal at temperatures ranging between 21° and 38°C and is retarded when the temperature falls below 10°C or when it exceeds 38°C. A high environmental humidity enhances putrefaction.[55,70,80–91]

Sign IX: Irreversible Loss of Brain Functions

As has already been emphasized, when physicians suspect an irreversible loss of brain function in a heart-beating and ventilatory supported case, BD diagnostic criteria are applied.

To assess brain functions during BD diagnosis, the clinical neurologic examination is the accepted standard (Table 4.1).[33]

Preconditions and Prerequisites

Coma Due to an Irreversible Acute Brain Damage of Known Etiology, Affecting Both Hemispheres and Brainstem

We must have clear and definitive clinical or neuroimaging evidence of an acute central nervous system (CNS) insult that is consistent with the irreversible loss of neurological function.[30,32,33,92–100]

The majority of comatose patients who finally evolve into BD have suffered severe head injury, aneurysmal subarachnoid hemorrhage, intracerebral hemorrhage, hypoxic-ischemic encephalopathy after prolonged cardiac resuscitation or asphyxia, large ischemic strokes associated with brain swelling and herniation, or massive brain edema in patients with fulminant hepatic necrosis (Fig. 4.3).[32,33,97,101]

It is essential to obtain a precise clinical history in suspected brain-dead patients, to determine a clear coma etiology, and to exclude those conditions that mimic BD. The introduction of computed tomography (CT) scan into medical practice has facilitated documenting brain damage and explaining the loss of brain and brainstem function. More recently, magnetic resonance imaging (MRI) has also provided a powerful tool for detecting an irreversible acute brain damage. In patients with normal CT or MRI findings, the clinical diagnosis of BD is doubtful. Nonetheless, some patients who have suffered an ischemic-anoxic insult show a normal CT and MRI. In these cases, some newly developed MRI functional techniques, such as diffusion weighted imaging, could be of great value.[102–113]

The presence of primary brainstem lesions in patients undergoing BD diagnosis warrants discussion. The loss of function of the brainstem is in a large majority of cases the infratentorial consequence of extremely severe supratentorial damage,

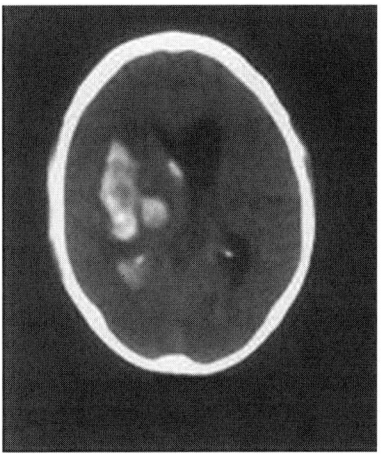

FIGURE 4.3. Computed tomography scan in a brain-dead patient showing an intracerebral hemorrhage.

resulting in untreatable intracranial pressure. Nevertheless, in some cases, primary brainstem lesions may entirely damage this structure while functioning of the cerebral hemispheres is relatively spared. This can be shown in a number of cases by the presence of quasi-normal EEG activity and the visual evoked potentials.[16,114–127] Hence, in primary brainstem lesions we have proposed the use of confirmatory ancillary tests (see Chapter 5).[128]

Confounding Factors to be Excluded

It is essential to exclude the confounding factors that mimic BD.[33,98,129]

Unresuscitated Shock

Neurological assessments may be unreliable in the acute post-resuscitation phase after cardiorespiratory arrest. Moreover, the presence of unstable blood pressure has also been reported in BD. Hence, it indispensable to apply BD criteria only with a systolic blood pressure minimum value of 90 mm Hg. The use of hydration fluids or vasopressor drugs might reestablish adequate blood pressure values.[33,92–95,101]

Hypothermia (Core Temperature <34°C)

An inability to regulate temperature, or poikilothermia, is often present in BD. Core temperature results should be obtained through central blood, rectal, or esophageal gastric measurement. Most sets of criteria for BD diagnosis require a body temperature of at least 32.2°C.[10,27,30,33,95,97,101,130] Recently, the Canadian Council for Donation and Transplantation stated that the 32.2°C standard in the context of severe brain injury is uncertain, and a pronouncement was made to adopt 34°C "as a rational, safe and attainable standard."[92]

Walker[13,25,30,131] reported that in the Collaborative Study, the body temperatures of only 12 of 503 patients were below 32°C. Milhaud et al.[132–134] had stated that a hypothermia below 32°C is one of the best signs of total brain death. Schneider and Matakas[135] compared the impaired thermoregulation in brain death to the reversible hypothermia that accompanies the spinal shock seen in high cervical cord lesions. Shewmon[136] has also compared spinal shock with BD.

The correction of hypothermia before applying criteria of BD is appropriate, especially in children and alcoholic patients who become hypothermic after mild brain injuries.[131] Moreover, it is important for physicians to be aware of subjects undergoing accidental hypothermia, as already discussed.

In contrast, Takasu et al.[137] stated that hyperthermia is a frequent finding during the post-resuscitation period after cardiac arrest. These authors also found that the incidence of a peak axillary temperature of 39°C during the first 72 hours of hospitalization was associated with a higher occurrence of BD. Hence, they concluded that hyperthermia at an early post-resuscitation stage after cardiac arrest is an indicator of a poor prognosis, highly correlated with an evolution to BD. Mild postischemic hyperthermia significantly exacerbates functional and structural neurologic injury after deep hypothermic circulatory arrest and therefore should be avoided. Other authors have also concluded that hyperthermia is an early indicator of brain damage after resuscitation.[137–141]

Severe Metabolic Disorders Capable of Causing a Potentially Reversible Coma

Severe metabolic or endocrine abnormalities make BD unreliable.[33,96–98,129] Metabolic or endocrine derangement including glucose, electrolytes (phosphate, calcium, and magnesium), inborn errors of metabolism, and liver or renal dysfunction may cause a potential reversible coma. If in the treating physician's judgment the metabolic abnormality may play a role, it should be corrected.[33]

Peripheral Nerve or Muscle Dysfunction

Peripheral nerve or muscle dysfunction and neuromuscular blockade potentially play a role in inducing unresponsiveness in patients, and it could be a confounding factor in BD diagnosis.[142–148]

Clinically Significant Drug Intoxications (e.g., Alcohol, Barbiturates, Sedatives, Hypnotics)

Central nervous system–depressant drugs should be ruled out if the clinical history is indicative. It is therefore mandatory that the drug history be carefully reviewed, and if there is any possibility of intoxication being the cause of, or contributing to, the patient's comatose state, it precludes a diagnosis of BD. In this case, drug screens are only supportive when tests are performed for a specific drug. Depressing drugs may have prolonged action, particularly when hypothermia coexists or in the context of renal or hepatic failure.[20,31,97,148–157] Nonetheless, therapeutic levels or therapeutic dosing of anticonvulsants, sedatives, and analgesics do not preclude the diagnosis.[92,94] Regarding barbiturates, BD diagnosis most likely can be done with subtherapeutic levels, even though reliable research in this area is lacking. In brain-dead neonates and children, an isoelectric EEG remains, even decrementing from therapeutic to subtherapeutic levels of barbiturates.[33,95,98]

Criteria

Deep Unresponsive Coma

The diagnosis of deep unresponsive coma requires that a comatose patient show a lack of spontaneous movements in addition to an absence of motor responses mediated by stimuli applied within the cranial nerve distribution.[11,17–20,30,33,38,101,131,158–161]

Of course, brain unresponsivity or inactivity diagnosis excludes all CNS functions related to consciousness (arousal and awareness), such as cognition, perception, speech, voluntary movements, and facial expressions. An unresponsive state of coma implies a complete lack of the above-mentioned CNS responses to any kind of external and internal excitation.[33,95,97,128]

A CNS-mediated motor response to pain in any distribution, seizures, and decorticate and decerebrate responses impede BD diagnosis. Some BD patients may present spinal reflexes or motor responses, confined to spinal distribution, which do not preclude the BD diagnosis.[33,95,97,101,131,162] In the original Harvard Committee Report, absent spinal reflexes was one of the diagnostic criteria.[2] Nonetheless, in the early 1970s, several publications appeared, documenting the preservation of spinal reflex activity in brain-dead patients.[29,163–168]

In BD, plantar responses, finger jerks, undulating toe flexion, muscle stretch reflexes, abdominal reflexes, triple flexion response, pronation–extension reflex, and facial myokymia have been described.[162,169–174]

Nonetheless, the most dramatic reflex movement is the Lazarus sign, named for Lazarus in the Bible. The Lazarus sign occurring in ICUs has frightened staff and family members, and is viewed as some kind of miracle or resuscitation.[175,176] Mandel et al.[177] reported a 28-year-old man with BD who spontaneously had extension movements of both upper limbs with hands held in a praying position. In general, the Lazarus sign is described as a sequence of complex movements, characterized by flexion of the arms at the elbow, adduction of the shoulders, lifting of the arms, dystonic posturing of the hands and fingers, and crossing of the hands; flexion/extension of the lower limbs as well as walking-like movements can occur. Most cases involve movements of the upper limbs only, although some also had lower limb motion, such as flexion at the hips and knees, or shallow and irregular respiration movements.[175–181]

Ropper[182] described the Lazarus sign as "unusual spontaneous movements in brain-dead patients" and reported it in five patients, precipitated after several minutes by hypoxia or ischemia when the ventilator was removed terminally, or during apnea testing. "The arms flex quickly to the chest from the patient's side, the shoulders adduct, and in some patients, the hands cross or oppose just below the chin. The limbs then return to the patient's side, sometimes asymmetrically." Other authors reported similar cases occurring during nonhypoxic conditions with associated arterial hypertension, tachycardia and facial flushing.[178,183] The

Lazarus sign was also reported upon either respiratory support removal or neck flexion.[175,176,179,180,184,185]

Jørgensen[186] emphasized that patients retained or regained spinal reflex activity after brain death. He introduced the term *spinal man* after brain death. He reported two main observations: (1) In individuals who lost spinal reflexes and then regained them, it occurred within 6 hours with the following progression: a flexion-withdrawal response first, and then cremasteric, abdominal, and finally deep tendon reflexes. (2) When spinal reflexes were delayed, it was invariably associated with severe arterial hypotension. Jørgensen also reported the presence of a cutaneously elicited pronation–extension reflex of the upper limbs in BD; the incidence of this reflex increased with clinical evolution time, from 5% (3 of 63 cases) within 6 to 12 hours, to 71% (17 of 24 cases) after 22 hours. Jørgensen concluded that "this reflex has proved to be a simple indicator of brain death of any aetiology."[174,186,187]

Ivan[165] retrospectively studied 52 patients with BD. He found deep tendon reflexes in 35%, plantar flexor responses in 60%, plantar withdrawal in 35%, and abdominal reflexes in 40% . He also observed neck-arm flexion in 25%, neck-hip flexion in 45%, and neck-abdominal reflex in 75%.

Saposnik et al.[162] have provided an important contribution in this area by developing prospective studies with standard protocols to elicit and record spinal reflexes. In their most recent publication on this topic, they reported a prospective study of 107 brain-dead patients. These authors in all cases followed a standardized protocol, especially designed to bring out motor movements, before and during an 8- to 10-minute apnea test, including (1) painful stimuli to the sternum, four limbs, and supraorbital area; (2) neck flexion; (3) tactile stimuli to the palmar and plantar areas; and (4) elevation of the four limbs. All patients with movements were recorded on video. Among 107 patients with a confirmed diagnosis of brain death, 47 (44%) had spontaneous or reflex movements. The most common movements were the undulating toe reflex and the triple flexion response. Twelve patients (26%) had more than one type of movement. It is interesting that these authors did not find statistically significant differences in age, sex, baseline temperature, blood pH, bicarbonate and partial pressure of carbon dioxide levels, and cause of death, comparing patients with and without movements. Moreover, patients with focal lesions who presented with greater frequency had spontaneous or reflex movements; those patients showing movements presented a higher mean blood pressure. Another important conclusion was that the cumulative proportion of patients showing movements increased with the time evolution after BD onset.[164] The authors described other motor movements, such as undulating toe flexion, triple flexion response, upper limb pronation–extension reflex, the Lazarus sign, and plantar flexor response.[162,169–173,175]

Other less frequent and bizarre reflex moments have been reported. Intermittent head turning from side to side or cephalic motor responses were seen when tactile or nociceptive stimuli were applied to the upper limbs.[188,189] Jastremski et al.[190] reported pseudodecerebrate postures affecting the four limbs, elicited by noxious stimuli or spontaneously. Marti-Fabregas et al.[191] reported pseudodecerebrate postures triggered by pulmonary mechanical insufflation or touch applied to arms, thorax, or abdomen in heart-beating cadavers. Santamaria et al.[192] described a BD patient with left eyelid opening and an unreactive pupil when noxious stimuli were applied to the ipsilateral nipple. Turnbull and Rutledge[193] reported an unusual rhythmic respiratory activity while the patient was on the ventilator, when continuous positive airway pressure ranged between 4 and 14 cm water. Nonetheless, Wijdicks et al.[194] recently noted false triggering in 5% of apnea tests. This happened when the ventilator settings were below a certain trigger sensitivity, causing the ventilator to self-cycle, simulating spontaneous respirations.

An infrequent movement that is difficult to explain in BD is facial myokymia. Saposnik et al. described a patient who had suffered a large putaminal hemorrhage and who fulfilled the BD criteria. The EEG was isoelectric. At that time, spontaneous twitches of the left cheek were noticed. They consisted of repetitive and rhythmic movements in groups of 3 ± 5 lasting for less than 5 seconds. These movements appeared every 2 ± 10 minutes for 6 hours. These authors proposed two possible peripheral mechanisms for this movement, such as muscle contraction as a result of denervation, or deafferentation of the facial nucleus (caused by supranuclear lesion). Hence, they concluded that facial myokymia in a brain-dead patient may represent a peripheral phenomenon and does not preclude the BD diagnosis.

Different pathophysiological mechanisms have been proposed to explain spinal reflexes in BD. Taking into consideration the Lazarus sign, it has been postulated that this movement sequence might be due to a stimulation of the cervical spinal cord, isolated from rostral brain areas, as has been reported during apnea testing or when arterial hypotension occurs.[178,179,181,195] The complex sequence of movements is explained by neuronal networks in the spinal cord, conforming to specific neural circuits, known as "central generators" in the spinal cord.[196]

As has already been discussed, noxious stimuli and passive neck flexion are triggering mechanisms, suggesting that stimuli corresponding to the cervical cord segments in BD might induce these reflexes.[162,168,170,172,173,175] Another possible mechanism is the spinal shock. Crenna et al.[197] monitored electrophysiologically and clinically the excitability of proprio- and exteroceptive spinal reflexes during the occurrence of BD. They concluded that the human spinal cord reacts to BD with a spinal shock, characterized by sequential recovery of reflex activity, and that this process is shorter than reflex recovery described in spinal shock following traumatic transection of the cord.

Shewmon[136] proposed that as the neuropathology of BD includes infarction down to the foramen magnum, the somatic pathophysiology of BD should resemble that of cervicomedullary junction transection plus vagotomy. Hence, he argued that the somatic pathophysiology of BD can be explained on the appearance of spinal reflexes—due to sudden cessation of rostral influences. Ibe[198] had also emphasized that primary spinal areflexia and the disorders of pulse and blood pressure that occurred after overcoming the spinal shock showed a clear tendency to recover.

These arguments demonstrate that spinal shock pathophysiology in BD might explain the appearance of spontaneous and reflex movements in brain-dead patients.[136]

The finding of these movements in BD has important practical and legal implications. The occurrence of spinal reflexes and movements in brain-dead patients may delay decision making, such as starting a transplantation procedure, because of difficulties in convincing the family or even a physician who is taking part in the diagnosis of brain death. Hence, it is important that the medical community should be aware that these movements can be present in brain-dead patients and do not invalidate the BD diagnosis.[162,173]

Absent Brainstem Reflexes

The reflexes mediated by the cranial nerves are the main indicators of the brainstem function. Hence, to prove their absence is an indispensable part of the BD diagnosis. The significance of each individual brainstem reflex varies in the BD diagnosis depending on the intrinsic sensitivity of the cranial nerve networks and on the effect of accompanying factors in BD patients, such as trauma, local edema, dried tissues, and intracranial tubes affecting the exploration of reflexes.[30,131]

Pupillary Reflex

The pupillary reflex, both direct and consensual, should be explored in both eyes using a bright light. This is an easy reflex to explore when the pupils are dilated, but if the pupils are small (2 to 3 mm), a convergent lens is recommended. This reflex is considered one of the most discriminant reflexes in BD diagnosis. It is also very helpful in evaluating prognosis in coma.[30,32,33,95,97,98,101,131] Pupil size is regulated by a balance of pupilloconstrictor (parasympathetic) and pupillodilator (sympathetic) reflexes.[131] The Harvard Criteria required that the pupil be fixed and dilated.[2] Mydriatic pupils can be found in BD, and the sympathetic cervical spine pathways connected to the pupillodilator muscle are intact. Nonetheless, several authors reported that mydriasis is not essential and that often small or medium-sized pupils are found in BD patients.[10,27,30,130] Although many drugs, such as glutethimide (Doriden) and scopolamine, affect pupil size, light reflex is very resistant and remains intact. Atropine in conventional doses, injected intravenously, has no important effect on pupil size. It also important to be aware of preexisting pupil abnormalities due to iris surgery and trauma to the cornea or bulbus oculi.[30,33,95,101,131,199-216]

Corneal Reflex

Unilateral corneal stimulation with a throat swab induces a bilateral closure of the eyelids, being an easy response to elicit, and if present, it is readily observed. This is also one of the most discriminant reflexes in BD diagnosis. A bilateral or unilateral response of eyelid closure and upward deviation of the eye (Bell's phenomenon) indicates preserved brainstem functioning. However, edema or drying of the cornea and severe facial and ocular trauma may preclude a satisfactory stimulus for this reflex. Moreover, the threshold for excitation decreases markedly if the lids are kept closed.[19,20,30,130,131,217−225]

Oculocephalic and Oculovestibular Responses

The oculocephalic reflex, also known as the doll's eyes response, is elicited upon brisk turning of the head from the centered position to 90 degrees on either side. In comatose patients without lesions of the brainstem, the eyes normally conjugately deviate to the other side. In BD no eye movements are observed. The neural pathways of this reflex are mediated through arcs involving the vestibular mechanisms, medial longitudinal bundle, and ocular nerves. This reflex is easy to obtain and is moderately discriminative. The oculovestibular response is elicited by irrigating the tympanum in both sides with ice water. In a comatose patient without lesions of the medial longitudinal bundle, or ocular nerves, the elicited response is a slow deviation of the eyes directed to the cold caloric stimulus. The head of the patient should be elevated 30 degrees above the horizontal plane, and 50 cc of ice water is irrigated into the external auditory canal using a small suction catheter.

To confirm lack of eye movement may be very tricky. It has been suggested that placing pen marks on the lower eyelid at the level of the pupil facilitates the detection of slight eye movements. The examiner should observe the patient for 1 minute after irrigation, and then wait an interval of 5 minutes before stimulating the opposite side. It is essential that the examiner confirm that the external auditory foramen and tympanic membrane are undamaged, because clotted blood or cerumen in the external auditory canal may impair this response. Moreover, a rupture of the eardrum usually augments caloric responses. There are also some drugs that can lessen this reflex, such as tricyclic antidepressants, aminoglycosides, antiepileptic drugs, anticholinergics, and chemotherapeutic agents. The presence of severe facial or ocular trauma, eyelid edema, and chemosis of the conjunctiva may limit movement of the globes, making it very difficult to elicit and observe eye movements. Basal fracture of the petrous bone abolishes the caloric response only unilaterally and may be identified by the presence of an ecchymotic mastoid process (Battle's sign).[6,27,30,32,33,38,96−98,101,131,158,159,162,188,201,226−230]

Gag and Cough Response

Pharyngeal (gag), cough, and swallowing reflexes are often difficult to explore because of the presence of tubes in the throat and dryness of the mucosa. Hence, in suspected BD patients the cough response is usually explored by passing a catheter through the endotracheal tube and suctioning with negative pressure for several seconds. As these reflexes have their arcs through the medulla oblongata, it is desirable to explore them.[27,30,33,231]

Atropine Test

The atropine test (AtT) assesses bulbar parasympathetic impact on heart activity in BD patients.[230] Jouvet[231] in 1959 proposed using atropine to explore bulbar activity for differentiating coma from death of the nervous system. Nonetheless,

Ouaknine[230,232−235] first proposed including this criterion for the so-called brainstem death. This test consists of injecting 2 mg of atropine under continuous monitoring with the electrocardiogram (ECG) for 10 minutes. The AtT is considered negative if the heart rate does not augment by more than 3% compared with basal ECG records. It should be applied after the patient meets other clinical BD criteria. There are, however, some limitations to the AtT. An infratentorial lesion or a brainstem lesion may cause damage to the dorsal vagal nucleus, resulting in a negative test result independent of BD. Moreover, the test is not applicable in cases of autonomic neuropathy and in patients who have undergone cardiac transplantation with denervation of autonomic fibers to the heart.

Nonetheless, as this test explores parasympathetic arcs through the medulla oblongata, and it is simple to apply, its inclusion is recommended in BD diagnostic protocols.[92,199,230,232−242] The atropine test is an efficient and simple diagnostic aid in cases of brain death.[34,128]

Apnea Test

The apnea test (AT) has been considered by some authors as the sine qua non for determining BD. It is an essential sign of a definitive loss of brainstem function, and implies that breathing has become extinct. Nonetheless, it is the most difficult, potentially harmful, and lengthy clinical test in BD protocols. Several reports have critiqued this test, emphasizing that it may induce hypotension, excessive hypercarbia, hypoxia, acidosis, cardiac arrhythmia, or asystole. Other authors defend it as the decisive clinical test in BD diagnosis.[19,20,93,133,194,195,243−277] Lang emphasized that an examiner with experience can apply this test in 99.9% of all cases; the remaining ones may be studied using other ancillary tests.[243,260,264,278−280]

Several methods have been proposed for monitoring blood gases, such as endtidal capnometry, pulse oximetry, transcutaneous blood gas determination, and intraarterial blood gas analysis,[260] although in vitro measurements, such as regular blood gas determinations, are considered the gold standard. Several authors propose determining a basal value and then repeating subsequent measurements every 5 minutes. Close monitoring is particularly valuable and indispensable when using artificial CO_2 augmentation.

According to Lang and Heckmann,[243,278] before applying this test some prerequisites should be considered:

- Most authors discourage clinicians from beginning this test at body temperatures below 32°C. Several authors recommend warming bodies to 36° or 36.5°C. Hypothermia correction facilitates CO_2 production, and reduces the chances of hypotension by stabilizing blood pressure and the hemodynamic state.[33,101,266,281]
- There are no clear PaO_2 recommendations except preoxygenation with 100% O_2 for some time, generally for 10 minutes, and avoidance of hypoxia. Preoxygenation of PaO_2 helps to correct possible hypocapnea, often due to hyperventilation or provoked by high tidal volumes of the mechanical ventilator, or due to hypothermia. The Quality Standards Subcommittee of the American Academy of

Neurology recommends a normal PO_2 or preoxygenation to obtain an arterial $PO_2 \geq 200$ mm Hg.

- The Quality Standards Subcommittee also recommends a normal arterial PCO_2 or $PaCO_2$ of ≥ 40 mm Hg.[17,101] Other authors suggest a starting $PaCO_2$ of 36 mm Hg or higher.[273,282] Telleria-Diaz[258] remarked that the main issue is to consider starting 20 mm Hg above the starting $PaCO_2$ level.
- A normal pH or a value in the low basic range is recommended before beginning AT.[247]
- A pretest systolic blood pressure (BP) of at least 90 mm Hg is recommended by the Quality Standards Subcommittee.[17,101]
- Fluid balance: euvolemia or a positive fluid balance during the previous 6 hours[101,243] is recommended.
- Medication: apnea testing must not be performed under the influence of drugs that can paralyze respiratory muscles, such as relaxants (e.g., pancuronium).

Procedure

Disconnection of Patient from Respirator. If continuous or intermittent oxygen supply is preceded by denitrogenation of blood gases, high PaO_2 levels can be sustained for very long periods of time.[276] Preoxygenation removes alveolar nitrogen stores and facilitates oxygen transport.[254] According to Lang and Heckmann,[243] there are techniques for ascertaining sufficient oxygenation during AT. The first technique is to disconnect the patient from the respirator and insert a catheter or cannula into the endotracheal tube down to the level of the carina, providing pure oxygen at a rate between 4 and 10 minutes. This ensures sufficient ventilation into the alveoli for transporting oxygen to the blood even without any respiratory movements.[33] In the second technique the patient is not disconnected from the respirator but the volume is decreased to a very low level (from 0.5 to 2 L/min), using synchronized intermittent mandatory volume ventilation and providing pure oxygen for inspiration; in this way, the patient is not disconnected until the required $PaCO_2$ is achieved. Lang prefers this method because it prevents tracheopulmonary complications and still allows the examiner to detect any spontaneous movement.[243,278] Al Jumah et al.[283] have also proposed using a biphasic intermittent positive airway pressure (BIPAP), a method knows as "bulk diffusion." Perel et al.[268,284] also considered it safer to test for apnea by leaving patients on a continuous flow of 100% oxygen and low peak end-expiratory pressure (PEEP) than to disconnect them from the ventilator.

Duration of Apnea Test. The apnea test result is positive, supporting a BD diagnosis, when no breathing effort is observed at a $PaCO_2$ of 60 mm Hg or with a 20 mm Hg increase from baseline. Since a rapid increase in $PaCO_2$ to 20 mm Hg above the normal baseline is considered a maximal stimulus to the respiratory centers, such an increase is recommended when the baseline $PaCO_2$ is at or above 36 to 40 mm Hg. In contrast, if respiratory movements are detected, the apnea test is classified as negative, not supporting a BD diagnosis, and the patient should be connected again to the respirator. If arterial pressure becomes <90 mm Hg, or a noticeable desaturation is detected by the oximeter, or cardiac arrhythmias take place, the procedure should be stopped and the ventilator should be connected. According to Lang, fixed durations are not considered standard anymore. Some authors have recommended 10 or 15 minutes, even if blood gas levels cannot be determined.[285,286] The rise in $PaCO_2$ predicted from normal physiology is 3 to 4 mm Hg,[287] but may vary considerably under conditions of brain death. Hence, Lang discourages such a procedure and insists on arterial blood gas determinations, because in his experience AT may last between 1 minute and more than 1 hour.

An alternative procedure has been proposed to perform standard apnea testing, insufflating CO_2 in cases where a decrease of ventilatory volume is not desired. Lang et al. reported that this method is mainly useful in cases showing a very slow rise of $PaCO_2$, or when blood oxygenation is hampered by pulmonary problems.[260,264,280] Careful and close monitoring must be ascertained, but if handled properly the test duration may be significantly shortened, providing a safer procedure in some cases.[243,278]

The level of PCO_2 has been proposed by several authors at which a maximal stimulation of the medullary respiratory centers occurs. Wijdicks has reported the occurrence of probably unsafe hypercarbia in some patients who have started to breathe at low PCO_2 levels of about 30 mm Hg; he described a patient who initiated breathing at PCO_2 levels around 35 mm Hg, and lost respiratory drive some hours later.[254,263] Wijdicks et al.[194] recently reported that when ventilator settings are below a certain trigger sensitivity threshold, some patients show false readings of spontaneous respiratory movements at rates of 20 to 30 per minute, which may cause confusion in BD determination.

Complications

Goudreau et al.[254] reported that approximately one in four apnea tests was associated with cardiovascular complications, and the rate of complications nearly doubled in tests without adequate preconditions. Inadequate preoxygenation had a clear association with an increased number of complications. Hypotension was the most frequent complication in 24% of cases, while cardiac arrhythmia with the potential for ventricular fibrillation or arrest was much less common (1%). Ropper et al.[285] reported that hypoxemia contributes to cardiac arrhythmia or hypotension during AT. Saposnik et al.[170] found complications in more that two thirds of their patients: hypotension (12%), acidosis (68%), and hypoxemia (23%). They reported that four patients suffered major complications, such as pneumothorax, cardiac arrest, bradycardia, atrial fibrillation, and myocardial infarction. Coimbra[288] directly critiqued the use of AT, reporting that it may induce potentially irreversible collapse of intracranial vessels by simultaneously worsening intracranial hypertension and inducing hypotension in BD patients. The author remarked that "apnea testing may induce rather than diagnose irreversible brain damage." This argument has been rejected by other authors, because "it has never been demonstrated that global cerebral ischemic penumbra is transformed into global brain infarction by

AT," suggesting future studies using serial positron emission tomography (PET) assessments.[247] The following are the main reported complications:

- *Excessive hypercarbia.* CO_2 values over 120 mm Hg cause narcosis and should be avoided, although general recommendations do not exist, and there are no generally accepted $PaCO_2$ values for patients whose natural respiration is adapted to a $PaCO_2$ of more than 45 mm Hg. In these cases other ancillary tests are recommended.[278]
- *Hypoxia.* O_2 values less than 70 or 60 mm Hg should be avoided. If pulse oximetry is used, values should not drop below 80%. It has been suggested that hypoxemia contributes to cardiac arrhythmia or hypotension during AT.[285]
- *Respiratory acidosis.* There are no precise recommendations for the pH, although pH values less than 7.2 or 7.0 should be avoided. Its fall is highly correlated with the increase of $PaCO_2$ and it is rapidly restored with normoventilation or mild hyperventilation. Some authors have published some low values during AT.[251,275] Lang and Heckmann et al.[243] have reported an acidosis as low as 6.808, and considered values down to 7.0 as acceptable; however, they recommended that pH should not be corrected by buffer or alkaline solutions, because alkalosis will result after normoventilation; as an alternative, a normal pH or a value in the low basic range should be established at the onset of AT.[243]
- *Hypotension.* Blood pressure values less than 60 or 70 mm Hg systolic, 40 mm Hg diastolic, and mean arterial pressure of 50 mm Hg should be avoided. The presence of a blood pressure of 90 mm Hg before testing is acceptable as a rule. Usually there is a mild increase of blood pressure with hyperoxygenation and a somewhat more marked decrease with hypercapnia. Persistent hypotension may be corrected using intravenous fluids, 5% albumin, or an increase of intravenous dopamine or (nor)epinephrine.
- *Cardiac arrhythmia and arrest.* Goudreau et al.[254] emphasized that cardiac arrhythmias induced by apnea testing are rare, and Saposnik et al.[170] reported cardiac arrest occurring only in two of 63 cases. It should be avoided at any rate. Arrhythmias are mostly due to excessive acidosis or to hypoxia, which can be suspected by the onset of new cardiac arrhythmias or a noticeable drop in heart rate.[243,278]
- *Barotrauma.* There have been several reports of the possibility of barotraumas,[247] although it is considered extremely rare.[170,195,289,290] Tension pneumothorax and pneumoperitoneum may ensue after massive oxygen insufflation into the endotracheal tube with an oxygen supply that obliterates the tube or a valve mechanism that makes the escape of the insufflated gas impossible.[195,247,290]
- *Increase of intracranial pressure.* Intracranial pressure (ICP) is usually not monitored during apnea testing except in some neurosurgical intensive care units. Since theoretically apnea testing may increase ICP via local hypercapnic vasodilation and the ensuing increase of cerebral blood volume, it should be the last of all clinical tests.[278]

Alternative Ancillary Tests

When a reliable AT is impossible to achieve, due to the unfeasibility of reaching the target $PaCO_2$, or when an unsafe drop in PaO_2 is inevitable, or in patients with severe thorax trauma or other pulmonary problems, other ancillary tests should be used (see Confirmatory Tests, below, and Chapter 5). This also could be recommended in those cases that were adapted to a $PaCO_2$ of more than 45 mm Hg.[243]

Periods of Observation

If BD determination is based only on clinical evaluation, the following periods of observation are proposed by the Cuban Commission[6,36]:

• 6 hours if a structural and irreversible CNS insult can be demonstrated by clinical and neuroimaging evidence
• 24 hours in cases of acute hypoxic-ischemic brain injury.

Periods of observation can be shortened according to medical criteria by the application of confirmatory tests (see Chapter 5).

Confirmatory Tests

Confirmatory tests are recommended to shorten the periods of observation according to the physicians' criteria and in those conditions interfering with the clinical BD diagnosis. The Cuban Commission proposed that the following confirmatory tests should be mandatory in patients with primary brainstem lesions undergoing BD diagnosis (see Chapter 5)[36,37,128]:

• Tests to demonstrate a complete cessation of brain circulation: TCD
• Confirmatory tests to demonstrate loss of bioelectrical activity: EEG, EP, and ERG.

References

1. Landmark article August 5, 1968: A definition of irreversible coma. Report of the Ad Hoc Committee of the Harvard Medical School to examine the definition of brain death. *JAMA.* 1984;252:677–679.
2. A definition of irreversible coma. Report of the Ad Hoc Committee of the Harvard Medical School to Examine the Definition of Brain Death. *JAMA.* 1968;205:337–340.
3. Stevens RD, Bhardwaj A. Approach to the comatose patient. *Crit Care Med.* 2006;34:31–41.
4. Bernat JL. The concept and practice of brain death. *Prog Brain Res.* 2005;150:369–379.
5. Munari M, Zucchetta P, Carollo C, et al. Confirmatory tests in the diagnosis of brain death: comparison between SPECT and contrast angiography. *Crit Care Med.* 2005;33:2068–2073.
6. Machado C. Determination of death. *Acta Anaesthesiol Scand.* 2005;49:592–593.

7. Young GB, Lee D. A critique of ancillary tests for brain death. *Neurocrit Care.* 2004;1:499–508.

8. Shiogai T, Saito I. [Ancillary studies in the determination of brain death]. *Nippon Rinsho.* 1997;55(suppl 1):301–309.

9. Pallis C. Further thoughts on brainstem death. *Anaesth Intensive Care.* 1995;23: 20–23.

10. Roosen K, Klein M. [Criteria and diagnosis of brain death]. *Versicherungsmedizin.* 1990;42:110–112.

11. Pallis C. Brainstem death: the evolution of a concept. *Semin Thorac Cardiovasc Surg.* 1990;2:135–152.

12. Gentleman D, Easton J, Jennett B. Brain death and organ donation in a neurosurgical unit: audit of recent practice. *BMJ.* 1990;301:1203–1206.

13. Walker AE. Brain death—an American viewpoint. *Neurosurg Rev.* 1989;12(suppl 1):259–264.

14. Galaske RG, Schober O, Heyer R. Determination of brain death in children with 123I-IMP and Tc-99m HMPAO. *Psychiatry Res.* 1989;29:343–345.

15. Pia HW. Brain death. *Acta Neurochir (Wien).* 1986;82:1–6.

16. Rodin E, Tahir S, Austin D, Andaya L. Brainstem death. *Clin Electroencephalogr.* 1985;16:63–71.

17. Korein J. The diagnosis of brain death. *Semin Neurol.* 1984;4:52–72.

18. Pallis CA, Prior PF. Guidelines for the determination of death. *Neurology.* 1983;33:251–252.

19. Pallis C. ABC of brain stem death. Diagnosis of brain stem death—II. *Br Med J (Clin Res Ed).* 1982;285:1641–1644.

20. Pallis C. ABC of brain stem death. Diagnosis of brain stem death—I. *Br Med J (Clin Res Ed).* 1982;285:1558–1560.

21. Pallis C. ABC of brain stem death. From brain death to brain stem death. *Br Med J (Clin Res Ed).* 1982;285:1487–1490.

22. Pallis C. ABC of brain stem death. Reappraising death. *Br Med J (Clin Res Ed).* 1982;285:1409–1412.

23. Pallis C, MacGillivary B. Brain death. *Lancet.* 1981;1:223.

24. Walker AE. Pathology of brain death. *Ann N Y Acad Sci.* 1978;315:272–280.

25. Walker AE. Ancillary studies in the diagnosis of brain death. *Ann N Y Acad Sci.* 1978;315:228–240.

26. Korein J. Brain death: interrelated medical and social issues. Terminology, definitions, and usage. *Ann N Y Acad Sci.* 1978;315:6–18.

27. An appraisal of the criteria of cerebral death. A summary statement. A collaborative study. *JAMA.* 1977;237:982–986.

28. Ingvar DH, Brun A, Johansson L, Sammuelsson SM. Survival after severe cerebral anoxia with destruction of the cerebral cortex: the apallic syndrome. In: Korein J, ed. *Brain Death: Interrelated Medical and Social Issues.* New York: Ann NY Acad Sci, 1977:184–214.

29. Walker AE, Diamond EL, Moseley J. The neuropathological findings in irreversible coma. A critique of the "respirator." *J Neuropathol Exp Neurol.* 1975;34:295–323.

30. Walker AE, Molinari GF. Criteria of cerebral death. *Trans Am Neurol Assoc.* 1975;100:29–35.

31. Ingvar DH, Widen L. [Brain death. Summary of a symposium]. *Lakartidningen.* 1972;69:3804–3814.

32. Wijdicks EFM. *Brain Death.* Philadelphia: Lippincott Williams & Wilkins, 2001.

33. Wijdicks EFM. Clinical diagnosis and confirmatory testing in brain death in adults. In: Wijdicks EFM, ed. *Brain Death*. Philadelphia: Lippincott Williams & Wilkins, 2001:61–90.

34. Machado C. Cerebral processing in the minimally conscious state. *Neurology.* 2005;65:973–974.

35. Machado-Curbelo C. [Do we defend a brain oriented view of death?]. *Rev Neurol.* 2002;35:387–396.

36. Machado C, Abeledo M, Alvarez C, et al. Cuba has passed a law for the determination and certification of death. *Adv Exp Med Biol.* 2004;550:139–142.

37. Machado C. [Resolution for the determination and certification of death in Cuba]. *Rev Neurol.* 2003;36:763–770.

38. Machado C, Sherman DL. *Brain Death and Disorders of Consciousness.* New York: Kluwer Academics/Plenum, 2004.

39. Schewe JC, Heister U, Fischer M, Hoeft A. [Accidental urban hypothermia. Severe hypothermia of 20.7 degrees C]. *Anaesthesist.* 2005;54:1005–1011.

40. Brat R, Skorpil J, Barta J, Suk M, Schichel T. Rewarming from severe accidental hypothermia with circulatory arrest. *Biomed Pap Med Fac Univ Palacky Olomouc Czech Repub.* 2004;148:51–53.

41. Roggla M, Frossard M, Wagner A, Holzer M, Bur A, Roggla G. Severe accidental hypothermia with or without hemodynamic instability: rewarming without the use of extracorporeal circulation. *Wien Klin Wochenschr.* 2002;114:315–320.

42. Francke A, Kopcke J. [Accidental hypothermia—a challenge for rescue service and intensive care]. *Anaesthesiol Reanim.* 2002;27:9–15.

43. Brat R, Suk M, Barta J, et al. [Resuscitation of a patient with deep hypothermia using extracorporeal circulation]. *Rozhl Chir.* 2002;81:279–281.

44. Farstad M, Andersen KS, Koller ME, Grong K, Segadal L, Husby P. Rewarming from accidental hypothermia by extracorporeal circulation. A retrospective study. *Eur J Cardiothorac Surg.* 2001;20:58–64.

45. Farbrot K, Husby P, Andersen KS, Grong K, Farstad M, Koller ME. [Rewarming of patients with accidental hypothermia with the help of heart-lung machine]. *Tidsskr Nor Laegeforen.* 2000;120:1854–1857.

46. Kilgus M, Simmen HP. Extracorporeal blood rewarming has proved to be a reliable method for treating patients suffering from accidental hypothermia (core temperature <28 degrees C). *World J Surg.* 2000;24:1282.

47. Gilbert M, Busund R, Skagseth A, Nilsen PA, Solbo JP. Resuscitation from accidental hypothermia of 13.7 degrees C with circulatory arrest. *Lancet.* 2000;355:375–376.

48. Brauer A, Wrigge H, Kersten J, Rathgeber J, Weyland W, Burchardi H. Severe accidental hypothermia: rewarming strategy using a veno-venous bypass system and a convective air warmer. *Intensive Care Med.* 1999;25:520–523.

49. Demaria R, Frapier JM, Valat J, et al. [Extracorporeal circulation for warming in severe accidental hypothermia. 3 cases]. *Presse Med.* 1998;27:664–666.

50. Cortes J, Galvan C, Sierra J, Franco A, Carceller J, Cid M. [Severe accidental hypothermia: rewarming by total cardiopulmonary bypass]. *Rev Esp Anestesiol Reanim.* 1994;41:109–112.

51. Machado C. Randomized clinical trial of magnesium, diazepam, or both after out-of-hospital cardiac arrest. *Neurology.* 2003;60:1868–1869.

52. Al-Alousi LM, Anderson RA, Worster DM, Land DV. Multiple-probe thermography for estimating the postmortem interval: I. Continuous monitoring and data analysis of

brain, liver, rectal and environmental temperatures in 117 forensic cases. *J Forensic Sci.* 2001;46:317–322.

53. Althaus L, Henssge C. Rectal temperature time of death nomogram: sudden change of ambient temperature. *Forensic Sci Int.* 1999;99:171–178.

54. Hooft P, van de Voorde H. The role of intestinal bacterial heat production in confounding postmortem temperature measurements. *Z Rechtsmed.* 1989;102:331–336.

55. Henssge C. [Estimation of death-time by computing the rectal body cooling under various cooling conditions (author's translation)]. *Z Rechtsmed.* 1981;87:147–178.

56. Moossa AR, Zarins CK, Skinner DB. In situ kidney preservation for transplantation with use of profound hypothermia 5 to 20 degrees C with an intact circulation. *Surgery.* 1976;79:60–64.

57. Bohnert M, Walther R, Roths T, Honerkamp J. A Monte Carlo-based model for steady-state diffuse reflectance spectrometry in human skin: estimation of carbon monoxide concentration in livor mortis. *Int J Legal Med.* 2005;119:355–362.

58. Bockholdt B, Maxeiner H, Hegenbarth W. Factors and circumstances influencing the development of hemorrhages in livor mortis. *Forensic Sci Int.* 2005;149:133–137.

59. Kaatsch HJ, Schmidtke E, Nietsch W. Photometric measurement of pressure-induced blanching of livor mortis as an aid to estimating time of death. Application of a new system for quantifying pressure-induced blanching in lividity. *Int J Legal Med.* 1994;106:209–214.

60. Kaatsch HJ, Stadler M, Nietert M. Photometric measurement of color changes in livor mortis as a function of pressure and time. Development of a computer-aided system for measuring pressure-induced blanching of livor mortis to estimate time of death. *Int J Legal Med.* 1993;106:91–97.

61. Hochmeister M, Dirnhofer R, Seifert D. [Can an intra vitam or postmortem restraint be diagnosed by the manifestation of livor mortis?]. *Beitr Gerichtl Med.* 1989;47:279–286.

62. Zagriadskaia AP, Makarov VI, Volodin SA, Fedorovtsev AL. [Detection of petechiae and livor mortis on putrefied and mummified corpses]. *Sud Med Ekspert.* 1987;30:30–32.

63. Schuller E, Pankratz H, Liebhardt E. [Colorimetry of livor mortis]. *Beitr Gerichtl Med.* 1987;45:169–173.

64. Fechner G, Koops E, Henssge C. [Cessation of livor in defined pressure conditions]. *Z Rechtsmed.* 1984;93:283–287.

65. Lins G. [Color location of livor mortis in spectrum analysis]. *Beitr Gerichtl Med.* 1973;31:203–212.

66. Cina SJ. Flow cytometric evaluation of DNA degradation: a predictor of postmortem interval? *Am J Forensic Med Pathol.* 1994;15:300–302.

67. Heller AR, Muller MP, Frank MD, Dressler J. [Rigor mortis—a definite sign of death?]. *Anasthesiol Intensivmed Notfallmed Schmerzther.* 2005;40:225–229.

68. Cavitt LC, Meullenet JF, Gandhapuneni RK, Youm GW, Owens CM. Rigor development and meat quality of large and small broilers and the use of Allo-Kramer shear, needle puncture, and razor blade shear to measure texture. *Poult Sci.* 2005;84:113–118.

69. Varetto L, Curto O. Long persistence of rigor mortis at constant low temperature. *Forensic Sci Int.* 2005;147:31–34.

70. Amendt J, Krettek R, Zehner R. Forensic entomology. *Naturwissenschaften.* 2004;91:51–65.

71. Gates J, Lam G, Ortiz JA, Losson R, Thummel CS. Rigor mortis encodes a novel nuclear receptor interacting protein required for ecdysone signaling during Drosophila larval development. *Development.* 2004;131:25–36.
72. Warriss PD, Brown SN, Knowles TG. Measurements of the degree of development of rigor mortis as an indicator of stress in slaughtered pigs. *Vet Rec.* 2003;153:739–742.
73. Hybrid rigor mortis. *Nat Biotechnol.* 2003;21:585.
74. Ratzliff AH, Santhakumar V, Howard A, Soltesz I. Mossy cells in epilepsy: rigor mortis or vigor mortis? *Trends Neurosci.* 2002;25:140–144.
75. Kobayashi M, Ikegaya H, Takase I, Hatanaka K, Sakurada K, Iwase H. Development of rigor mortis is not affected by muscle volume. *Forensic Sci Int.* 2001;117:213–219.
76. Alvarado CZ, Sams AR. Rigor mortis development in turkey breast muscle and the effect of electrical stunning. *Poult Sci.* 2000;79:1694–1698.
77. Alvarado CZ, Sams AR. The influence of postmortem electrical stimulation on rigor mortis development, calpastatin activity, and tenderness in broiler and duck pectoralis. *Poult Sci.* 2000;79:1364–1368.
78. Schwartz LM, Ruff RL. Changes in contractile properties of skeletal muscle during developmentally programmed atrophy and death. *Am J Physiol Cell Physiol.* 2002;282:C1270–C1277.
79. Pernick MS. Back from the grave: recurring controversies over defining and diagnosing death in history. In: Zaner RM, ed. *Death: Beyond the Whole Brain Criteria.* New York: Kluwer Academic, 1988:17–74.
80. Rothschild MA, Schmidt V, Pedal I. [Cadaver lipid: various origins add to the difficulty of assessing postmortem time]. *Arch Kriminol.* 1996;197:165–174.
81. Prieto JL, Magana C, Ubelaker DH. Interpretation of postmortem change in cadavers in Spain. *J Forensic Sci.* 2004;49:918–923.
82. Elliott S, Lowe P, Symonds A. The possible influence of micro-organisms and putrefaction in the production of GHB in post-mortem biological fluid. *Forensic Sci Int.* 2004;139:183–190.
83. Archer MS. Rainfall and temperature effects on the decomposition rate of exposed neonatal remains. *Sci Justice.* 2004;44:35–41.
84. Fiedler S, Graw M. Decomposition of buried corpses, with special reference to the formation of adipocere. *Naturwissenschaften.* 2003;90:291–300.
85. Sannohe S. Change in the postmortem formation of hypostasis in skin preparations 100 micrometers thick. *Am J Forensic Med Pathol.* 2002;23:349–354.
86. Whittaker DK, Rawle L. The effects of putrefaction on haptoglobin phenotyping from tooth fragments. *Forensic Sci Int.* 1994;64:151–157.
87. Payen G, Rimoux L, Gueux M, Lery N. [Body putrefaction in "air tight" burials. IV. Microbiologic findings and environment]. *Acta Med Leg Soc (Liege).* 1988;38:153–163.
88. Payen G, Lery N, Rimoux L, Gueux M, Ardouin R. [Body putrefaction in air-tight burials. III. Macroscopic changes of cadavers and contamination flow]. *Acta Med Leg Soc (Liege).* 1988;38:145–151.
89. Lery N, Isnard E, Payen G, Rimoux L, Gueux M. [Body putrefaction in air-tight burials. I. Legislative aspect and synthesis of acquired scientific knowledge]. *Acta Med Leg Soc (Liege).* 1988;38:115–131.
90. Morovic-Budak A. Experiences in the process of putrefaction in corpses buried in earth. *Med Sci Law.* 1965;18:40–43.
91. Trojanowska M. [Studies on the effect of putrefaction on the recovery of alkaloids from corpses.]. *Acta Pol Pharm.* 1962;19:453–458.

92. Guidelines for the diagnosis of brain death. Canadian Neurocritical Care Group. *Can J Neurol Sci.* 1999;26:64–66.
93. Parker BL, Frewen TC, Levin SD, et al. Declaring pediatric brain death: current practice in a Canadian pediatric critical care unit. *Can Med Assoc J.* 1995;153:909–916.
94. Death and brain death: a new formulation for Canadian medicine. Canadian Congress Committee on Brain Death. *Can Med Assoc J.* 1988;138:405–406.
95. Wijdicks EF. The first organ transplant from a brain-dead donor. *Neurology.* 2006;66:460–461.
96. Wijdicks EF, Cranford RE. Clinical diagnosis of prolonged states of impaired consciousness in adults. *Mayo Clin Proc.* 2005;80:1037–1046.
97. Wijdicks EF. The diagnosis of brain death. *N Engl J Med.* 2001;344:1215–1221.
98. Ashwal S. Clinical diagnosis and confirmatory testing of brain death in children. In: Wijdicks EFM, ed. *Brain Death.* Philadelphia: Lippincott Williams & Wilkins, 2001:91–114.
99. Walker AE. Advances in the determination of cerebral death. *Adv Neurol.* 1979;22:167–177.
100. Moseley JI, Molinari GF, Walker AE. Respirator brain. Report of a survey and review of current concepts. *Arch Pathol Lab Med.* 1976;100:61–64.
101. Wijdicks EF. Determining brain death in adults. *Neurology.* 1995;45:1003–1011.
102. Sener RN. Diffusion MRI in the postmortem brain: case report. *J Neuroradiol.* 2004;31:406–408.
103. Karantanas AH, Hadjigeorgiou GM, Paterakis K, Sfiras D, Komnos A. Contribution of MRI and MR angiography in early diagnosis of brain death. *Eur Radiol.* 2002;12:2710–2716.
104. Turner R, Le BD, Moonen CT, Despres D, Frank J. Echo-planar time course MRI of cat brain oxygenation changes. *Magn Reson Med.* 1991;22:159–166.
105. Schiff ND, Rodriguez-Moreno D, Kamal A, et al. fMRI reveals large-scale network activation in minimally conscious patients. *Neurology.* 2005;64:514–523.
106. Wijdicks EF, Campeau NG, Miller GM. MR imaging in comatose survivors of cardiac resuscitation. *AJNR Am J Neuroradiol.* 2001;22:1561–1565.
107. Nakahara M, Ericson K, Bellander BM. Diffusion-weighted MR and apparent diffusion coefficient in the evaluation of severe brain injury. *Acta Radiol.* 2001;42:365–369.
108. Clawson TF, Vannucci SJ, Wang GM, Seaman LB, Yang XL, Lee WH. Hypoxia-ischemia-induced apoptotic cell death correlates with IGF-I mRNA decrease in neonatal rat brain. *Biol Signals Recept.* 1999;8:281–293.
109. Ishii K, Onuma T, Kinoshita T, Shiina G, Kameyama M, Shimosegawa Y. Brain death: MR and MR angiography. *AJNR Am J Neuroradiol.* 1996;17:731–735.
110. Orrison WW Jr, Champlin AM, Kesterson OL, Hartshorne MF, King JN. MR 'hot nose sign' and 'intravascular enhancement sign' in brain death. *AJNR Am J Neuroradiol.* 1994;15:913–916.
111. Jones KM, Barnes PD. MR diagnosis of brain death. *AJNR Am J Neuroradiol.* 1992;13:65–66.
112. Phan TG, Wijdicks EF. Diffusion-weighted magnetic resonance imaging in brain death. *Stroke.* 2000;31:1458–1459.
113. Lovblad KO, Bassetti C. Diffusion-weighted magnetic resonance imaging in brain death. *Stroke.* 2000;31:539–542.
114. Biniek R, Ferbert A, Buchner H, Bruckmann H. Loss of brainstem acoustic evoked potentials with spontaneous breathing in a patient with supratentorial lesion. *Eur Neurol.* 1990;30:38–41.

115. Buchner H, Schuchardt V. Reliability of electroencephalogram in the diagnosis of brain death. *Eur Neurol.* 1990;30:138–141.

116. Ogata J, Imakita M, Yutani C, Miyamoto S, Kikuchi H. Primary brainstem death: a clinico-pathological study. *J Neurol Neurosurg Psychiatry.* 1988;51:646–650.

117. Link J, Wagner W, Rohling R, Muhlberg J. [Is cerebral panangiography unnecessary in determining brain death?]. *Anaesthesist.* 1988;37:43–48.

118. Dubikaitis I, Borshchagovskii ML, Poliakova VB, Tigliev GS. [Clinico-physiologic evaluation of the viability of the brain as an organ in segmental lesions of the brain stem]. *Zh Vopr Neirokhir Im N N Burdenko.* 1983;19–23.

119. Zeitlhofer J, Mayr N, Auff E. [The value of evoked potentials in brain death diagnosis]. *Wien Med Wochenschr.* 1990;140:564–567.

120. Powner DJ, Fromm GH. The electroencephalogram in the determination of brain death. *N Engl J Med.* 1979;300:502.

121. Ashwal S, Schneider S. Failure of electroencephalography to diagnose brain death in comatose children. *Ann Neurol.* 1979;6:512–517.

122. Wilkus RJ, Harvey F, Ojemann LM, Lettich E. Electroencephalogram and sensory evoked potentials. Findings in an unresponsive patient with pontine infarct. *Arch Neurol.* 1971;24:538–544.

123. Grigg MM, Kelly MA, Celesia GG, Ghobrial MW, Ross ER. Electroencephalographic activity after brain death. *Arch Neurol.* 1987;44:948–954.

124. Schneider D. [Limits of resuscitation. II. Cerebral death and declaring the patient dead]. *Z Gesamte Inn Med.* 1981;36:421–424.

125. Schneider D, Baumann I, Kohler H. [Value and worthlessness of clinical studies in the diagnosis of brain death]. *Z Gesamte Inn Med.* 1976;31:29–37.

126. Darby J, Yonas H, Brenner RP. Brainstem death with persistent EEG activity: evaluation by xenon-enhanced computed tomography. *Crit Care Med.* 1987;15:519–521.

127. Ferbert A, Buchner H, Ringelstein EB, Hacke W. Isolated brain-stem death. Case report with demonstration of preserved visual evoked potentials (VEPs). *Electroencephalogr Clin Neurophysiol.* 1986;65:157–160.

128. Machado C, García OD. Guidelines for the determination of brain death. In: Machado C, ed. *Brain Death (Proceedings of the Second International Symposium on Brain Death).* Amsterdam: Elsevier Science BV, 1995:75–80.

129. de Tourtchaninoff M, Hantson P, Mahieu P, Guérit JM. Brain death diagnosis in misleading conditions. *Q J Med.* 1999;92:407–414.

130. Criteria for the diagnosis of brain stem death. Review by a working group convened by the Royal College of Physicians and endorsed by the Conference of Medical Royal Colleges and their Faculties in the United Kingdom. *J R Coll Physicians Lond.* 1995;29:381–382.

131. Walker AE. *Cerebral Death,* 3rd ed. Baltimore-Munich: Urban & Schwarzenberg, 1985.

132. Milhaud A, Groshens A. [Brain death]. *Agressologie.* 1986;27:957–958.

133. Milhaud A, Riboulot M, Gayet H. Disconnecting tests and oxygen uptake in the diagnosis of total brain death. *Ann N Y Acad Sci.* 1978;315:241–251.

134. Milhaud A, Ossart M, Gayet H, Riboulot M. [The oxygen disconnecting test. A test for brain death]. *Ann Anesthesiol Fr.* 1974;15(spec. no. 3):73–79.

135. Schneider H, Matakas F. Pathological changes of the spinal cord after brain death. *Acta Neuropathol (Berl).* 1971;18:234–247.

136. Shewmon DA. Spinal shock and 'brain death': somatic pathophysiological equivalence and implications for the integrative-unity rationale. *Spinal Cord.* 1999;37:313–324.

137. Takasu A, Saitoh D, Kaneko N, Sakamoto T, Okada Y. Hyperthermia: is it an ominous sign after cardiac arrest? *Resuscitation.* 2001;49:273–277.

138. del Castillo Fernandez de Betono, Cruz PP, Garutti I, Fernandez-Quero L, Ferrando A, Munoz J. [Brain death and hyperthermia]. *Rev Esp Anestesiol Reanim.* 2002;49:435–436.

139. Shum-Tim D, Nagashima M, Shinoka T, et al. Postischemic hyperthermia exacerbates neurologic injury after deep hypothermic circulatory arrest. *J Thorac Cardiovasc Surg.* 1998;116:780–792.

140. Takino M, Okada Y. Hyperthermia following cardiopulmonary resuscitation. *Intensive Care Med.* 1991;17:419–420.

141. Lausberg G. [Central disorders of temperature regulation. Clinical-experimental study]. *Acta Neurochir (Wien).* 1972;19(suppl):168.

142. Hansen PB, Thisted BK, Andersen LW, Larsen B, Jansen EC, Secher NH. [Anesthesiologic aspects of multiorgan donation]. *Ugeskr Laeger.* 1992;154:1172–1176.

143. Vargas F, Hilbert G, Gruson D, Valentino R, Gbikpi-Benissan G, Cardinaud JP. Fulminant Guillain-Barre syndrome mimicking cerebral death: case report and literature review. *Intensive Care Med.* 2000;26:623–627.

144. Shewmon DA. Recovery from "brain death": a neurologist's apologia. *Linacre Q.* 1997;64:30–96.

145. Bakshi N, Maselli RA, Gospe SM Jr, Ellis WG, McDonald C, Mandler RN. Fulminant demyelinating neuropathy mimicking cerebral death. *Muscle Nerve.* 1997;20:1595–1597.

146. Bohlega SA, Stigsby B, Haider A, McLean D. Guillain-Barre syndrome with severe demyelination mimicking axonopathy. *Muscle Nerve.* 1997;20:514–516.

147. Coad NR, Byrne AJ. Guillain-Barre syndrome mimicking brainstem death. *Anaesthesia.* 1990;45:456–457.

148. Ostermann ME, Young B, Sibbald WJ, Nicolle MW. Coma mimicking brain death following baclofen overdose. *Intensive Care Med.* 2000;26:1144–1146.

149. Brenner RP. The interpretation of the EEG in stupor and coma. *Neurologist.* 2005;11:271–284.

150. Marrache F, Megarbane B, Pirnay S, Rhaoui A, Thuong M. Difficulties in assessing brain death in a case of benzodiazepine poisoning with persistent cerebral blood flow. *Hum Exp Toxicol.* 2004;23:503–505.

151. Sztajnkrycer MD, Huang EE, Bond GR. Acute zonisamide overdose: a death revisited. *Vet Hum Toxicol.* 2003;45:154–156.

152. Caballero F, Lopez-Navidad A, Cotorruelo J, Txoperena G. Ecstasy-induced brain death and acute hepatocellular failure: multiorgan donor and liver transplantation. *Transplantation.* 2002;74:532–537.

153. Jones AL, Simpson KJ. Drug abusers and poisoned patients: a potential source of organs for transplantation? *Q J Med.* 1998;91:589–592.

154. Flanagan RJ. The poisoned patient: the role of the laboratory. *Br J Biomed Sci.* 1995;52:202–213.

155. Kosteljanetz M. [Drug poisoning vs. brain death]. *Ugeskr Laeger.* 1991;153:2244.

156. Thompson AE, Sussmane JB. Bretylium intoxication resembling clinical brain death. *Crit Care Med.* 1989;17:194–195.

157. Blake KV, Bailey D, Zientek GM, Hendeles L. Death of a child associated with multiple overdoses of acetaminophen. *Clin Pharm.* 1988;7:391–397.
158. Truog RD, Flacker JC. Rethinking brain death. *Crit Care Med.* 2006;20:1705–1713.
159. Michelson DJ, Ashwal S. Evaluation of coma and brain death. *Semin Pediatr Neurol.* 2004;11:105–118.
160. Machado-Curbelo C. [A new formulation of death: definition, criteria and diagnostic tests]. *Rev Neurol.* 1998;26:1040–1047.
161. Bernat JL. The definition, criterion, and statute of death. *Semin Neurol.* 1984;4:45–51.
162. Saposnik G, Maurino J, Saizar R, Bueri JA. Spontaneous and reflex movements in 107 patients with brain death. *Am J Med.* 2005;118:311–314.
163. de Lecea L, Sutcliffe JG. The hypocretins and sleep. *FEBS J.* 2005;272:5675–5688.
164. Robert F, Mumenthaler M. [Criteria of brain death. Spinal reflexes in 45 personal studies]. *Schweiz Med Wochenschr.* 1977;107:335–341.
165. Ivan LP. Spinal reflexes in cerebral death. *Neurology.* 1973;23:650–652.
166. Schneider H, Matakas F, Simon RS, Cervos-Navarro J. [Intracranial hypertension and total ischemic infarction of the brain]. *Neurochirurgie.* 1972;18:159–170.
167. Schneider H. [Brain death and moelle isolee]. *Dtsch Med Wochenschr.* 1970;95:1538.
168. Butenuth J, Schneider H, Schneider V. [Spinal mechanisms in a case of cerebral death after cyanide poisoning]. *Dtsch Z Nervenheilkd.* 1970;197:255–284.
169. Saposnik G, Maurino J, Saizar R, Bueri JA. Undulating toe movements in brain death. *Eur J Neurol.* 2004;11:723–727.
170. Saposnik G, Rizzo G, Vega A, Sabbatiello R, Deluca JL. Problems associated with the apnea test in the diagnosis of brain death. *Neurol India.* 2004;52:342–345.
171. Saposnik G, Maurino J, Saizar R. Facial myokymia in brain death. *Eur J Neurol.* 2001;8:227–230.
172. Saposnik G, Maurino J, Bueri J. Movements in brain death. *Eur J Neurol.* 2001;8:209–213.
173. Saposnik G, Bueri JA, Maurino J, Saizar R, Garretto NS. Spontaneous and reflex movements in brain death. *Neurology.* 2000;54:221–223.
174. Jørgensen EO. Spinal man after brain death. In: Machado C, ed. *Brain Death (Proceedings of the Second International Symposium on Brain Death).* Amsterdam: Elsevier Science BV, 1995:87–94.
175. Bueri JA, Saposnik G, Maurino J, Saizar R, Garretto NS. Lazarus' sign in brain death. *Mov Disord.* 2000;15:583–586.
176. Urasaki E, Fukumura A, Itho Y, et al. [Lazarus' sign and respiratory-like movement in a patient with brain death]. *No To Shinkei.* 1988;40:1111–1116.
177. Mandel S, Arenas A, Scasta D. Spinal automatism in cerebral death. *N Engl J Med.* 1982;307:501.
178. Heytens L, Verlooy J, Gheuens J, Bossaert L. Lazarus sign and extensor posturing in a brain-dead patient. Case report. *J Neurosurg.* 1989;71:449–451.
179. Awada A. [Uncommon reflex automatisms after brain death]. *Rev Neurol (Paris).* 1995;151:586–588.
180. Nokura K, Yamamoto H, Uchida M, Hashizume Y, Inagaki T. [Automatic movements of extremities induced in primary massive brain lesion with apneic coma]. *Rinsho Shinkeigaku.* 1997;37:198–207.
181. Turmel A, Roux A, Bojanowski MW. Spinal man after declaration of brain death. *Neurosurgery.* 1991;28:298–301.
182. Ropper AH. Unusual spontaneous movements in brain-dead patients. *Neurology.* 1984;34:1089–1092.

183. Jordan JE, Dyess E, Cliett J. Unusual spontaneous movements in brain-dead patients. *Neurology.* 1985;35:1082.

184. Urasaki E, Tokimura T, Kumai J, Wada S, Yokota A. Preserved spinal dorsal horn potentials in a brain-dead patient with Lazarus' sign. Case report. *J Neurosurg.* 1992;76:710–713.

185. Dequin PF, Hazouard E, Huguet M, Delplace C, Tayoro J, Ginies G. [Sign of Lazarus after brain death]. *Presse Med.* 1998;27:1102.

186. Jørgensen EO. Spinal man after brain death. The unilateral extension-pronation reflex of the upper limb as an indication of brain death. *Acta Neurochir (Wien).* 1973;28:259–273.

187. Holm S, Jørgensen EO. Ethical issues in cardiopulmonary resuscitation. *Resuscitation.* 2001;50:135–139.

188. Christie JM, O'Lenic TD, Cane RD. Head turning in brain death. *J Clin Anesth.* 1996;8:141–143.

189. Jokelainen M, Marttila R. Cephalic motor responses in brain death. *Br Med J.* 1977;2:828.

190. Jastremski MS, Powner D, Snyder J, Smith J, Grenvik A. Spontaneous decerebrate movement after declaration of brain death. *Neurosurgery.* 1991;29:479–480.

191. Marti-Fabregas J, Lopez-Navidad A, Caballero F, Otermin P. Decerebrate-like posturing with mechanical ventilation in brain death. *Neurology.* 2000;54:224–227.

192. Santamaria J, Orteu N, Iranzo A, Tolosa E. Eye opening in brain death. *J Neurol.* 1999;246:720–722.

193. Turnbull J, Rutledge F. Spontaneous respiratory movements with clinical brain death. *Neurology.* 1985;35:1260.

194. Wijdicks EF, Manno EM, Holets SR. Ventilator self-cycling may falsely suggest patient effort during brain death determination. *Neurology.* 2005;65:774.

195. Lessard M, Mallais R, Turmel A. Apnea test in the diagnosis of brain death. *Can J Neurol Sci.* 2000;27:353–354.

196. Dietz V. Human neuronal control of automatic functional movements: interaction between central programs and afferent input. *Physiol Rev.* 1992;72:33–69.

197. Crenna P, Conci F, Boselli L. Changes in spinal reflex excitability in brain-dead humans. *Electroencephalogr Clin Neurophysiol.* 1989;73:206–214.

198. Ibe K. Clinical and pathophysiological aspects of the intravital brain death [abstract]. *Electroencephalogr Clin Neurophysiol.* 1971;30(3):272.

199. Zhang TX, Shi YJ, Sheng HQ, Wei GH, Zhao WG. [Clinical diagnosis of 26 cases of brain death]. *Zhonghua Yi Xue Za Zhi.* 2004;84:93–96.

200. Shlugman D, Parulekar M, Elston JS, Farmery A. Abnormal pupillary activity in a brainstem-dead patient. *Br J Anaesth.* 2001;86:717–720.

201. Kennedy MC, Moran JL, Fearnside M, et al. Drugs and brain death. *Med J Aust.* 1996;165:394–398.

202. Larson MD, Muhiudeen I. Pupillometric analysis of the 'absent light reflex'. *Arch Neurol.* 1995;52:369–372.

203. Shirakawa S, Ishikawa S. [Evaluation of autonomic nervous function by pupil dynamics recording]. *Nippon Rinsho.* 1992;50:708–716.

204. Ishiguro T, Tamagawa S, Ogawa H. [Changes of pupil size in brain death patients]. *Seishin Shinkeigaku Zasshi.* 1992;94:864–873.

205. Ishikawa S. [Pupil—recent development of investigation]. *Rinsho Shinkeigaku.* 1990;30:1329–1333.

206. Bleck TP. Dilated pupils and brain death. *Ann Intern Med.* 1990;112:632–633.

207. Fogelholm R, Laru-Sompa R. Brain death and pinpoint pupils. *J Neurol Neurosurg Psychiatry.* 1988;51:1002.
208. Larson MD. Action of narcotics on the pupil. *Anaesthesia.* 1987;42:566.
209. Verma NP. Drugs as a cause of fixed, dilated pupils after resuscitation. *JAMA.* 1986;255:3251.
210. Pallis C. ABC of brain stem death. Pitfalls and safeguards. *Br Med J (Clin Res Ed).* 1982;285:1720–1722.
211. Allen N, Burkholder J, Comiscioni J. Clinical criteria of brain death. *Ann N Y Acad Sci.* 1978;315:70–96.
212. Black PM. Brain death (first of two parts). *N Engl J Med.* 1978;299:338–344.
213. Safran AB, Werner A, Babel J. [Ophthalmodynamometry: a complementary method in the diagnosis of irreversible coma]. *Schweiz Arch Neurol Neurochir Psychiatr.* 1977;120:43–49.
214. Sims JK, Bickford RG. Non-mydriatic pupils occurring in human brain death. *Bull Los Angeles Neurol Soc.* 1973;38:24–32.
215. Wexler J. Pupil size and irreversible coma. *N Engl J Med.* 1971;285:526.
216. Sims JK. Pupillary diameter in irreversible coma. *N Engl J Med.* 1971;285:57.
217. Pallis C. Further thoughts on brainstem death. *Anaesth Intensive Care.* 1995;23:20–23.
218. Walker AE. The death of a brain. *Johns Hopkins Med J.* 1969;124:190–201.
219. Giacino JT. The vegetative and minimally conscious states: consensus-based criteria for establishing diagnosis and prognosis. *NeuroRehabilitation.* 2004;19:293–298.
220. Taylor RM. Reexamining the definition and criteria of death. *Semin Neurol.* 1997;17:265–270.
221. Jørgensen EO, Brodersen P. [Criteria of death. The value of cliniconeurologic examination of cerebral circulation for the assessment of brain death]. *Nord Med.* 1971;86:1549–1560.
222. O'Leary DJ, Millodot M. Brain death and the corneal reflex. *Lancet.* 1980;2:1379.
223. Searle J, Collins C. A brain-death protocol. *Lancet.* 1980;1:641–643.
224. Molnar L, Soos A. [The ophthalmological symptoms of cerebral and biological death (author's translation)]. *Klin Monatsbl Augenheilkd.* 1974;165:828–831.
225. Park DM. Corneo-retinal potentials in severe brain damage. *Lancet.* 1970;2:933–934.
226. Wijdicks EFM. *Neurological Complications of Critical Illness,* 2nd ed. Oxford: Oxford University Press, 2002.
227. Wijdicks EF. Neurologic complications in critically ill patients. *Anesth Analg.* 1996;83:411–419.
228. Larson MD, Gray AT. The diagnosis of brain death. *N Engl J Med.* 2001;345:616–617.
229. Walker AE. Dead or alive. *J Nerv Ment Dis.* 1984;172:639–641.
230. Drory Y, Ouaknine G, Kosary IZ, Kellermann JJ. Electrocardiographic findings in brain death; description and presumed mechanism. *Chest.* 1975;67:425–432.
231. Jouvet M. Diagnostic électro-sous-cortico-graphique de la mort du système nerveux central au cours de certains comas. *Electroencephalogr Clin Neurophysiol.* 1959;11(4):805–808.
232. Ouaknine GE, Mercier C. [Value of the atropine test in the confirmation of brain death]. *Union Med Can.* 1985;114:76–80.
233. Ouaknine GE. Cardiac and metabolic alterations in brain death: discussion paper. *Ann N Y Acad Sci.* 1978;315:252–264.

234. Ouaknine G, Kosary IZ, Braham J, Czerniak P, Nathan H. Laboratory criteria of brain death. *J Neurosurg.* 1973;39:429–433.

235. Ouaknine G, Kosary IZ, Ziv M. [Value of the caloric test and electronystagmography in the diagnosis of irreversible coma]. *Neurochirurgie.* 1973;19:407–414.

236. Su YY, Zhao H, Zhang Y, Wang XM, Hua Y. [Studies on evaluation of brain death]. *Zhonghua Nei Ke Za Zhi.* 2004;43:250–253.

237. Huttemann E, Schelenz C, Sakka SG, Reinhart K. Atropine test and circulatory arrest in the fossa posterior assessed by transcranial Doppler. *Intensive Care Med.* 2000;26:422–425.

238. Goetting MG, Contreras E. Systemic atropine administration during cardiac arrest does not cause fixed and dilated pupils. *Ann Emerg Med.* 1991;20:55–57.

239. Siemens P, Hilger HH, Frowein RA. Heart rate variability and the reaction of heart rate to atropine in brain dead patients. *Neurosurg Rev.* 1989;12 Suppl 1:282–284.

240. Vaghadia H. Atropine resistance in brain-dead organ donors. *Anesthesiology.* 1986;65:711–712.

241. Cardan C, Roth A, Biro J. [The atropine test in the assessment of brain death]. *Rev Chir Oncol Radiol O R L Oftalmol Stomatol Chir.* 1983;32:393–397.

242. Jørgensen EO. Aortocervical, internal carotid and vertebral angiography in total cerebral infarction. *Acta Radiol Diagn (Stockh).* 1973;14:369–378.

243. Lang CJ, Heckmann JG. Apnea testing for the diagnosis of brain death. *Acta Neurol Scand.* 2005;112:358–369.

244. Tsai WH, Lee WT, Hung KL. Determination of brain death in children—a medical center experience. *Acta Paediatr Taiwan.* 2005;46:132–137.

245. Burns JM, Login IS. Confounding factors in diagnosing brain death: a case report. *BMC Neurol.* 2002;2:5.

246. Melano R, Adum ME, Scarlatti A, Bazzano R, Araujo JL. Apnea test in diagnosis of brain death: comparison of two methods and analysis of complications. *Transplant Proc.* 2002;34:11–12.

247. Vivien B, Haralambo MS, Riou B. [Barotrauma during apnea testing for the determination of brain death]. *Ann Fr Anesth Reanim.* 2001;20:370–373.

248. Jeret JS. Complications during apnea testing in the determination of brain death: predisposing factors. *Neurology.* 2001;56:1249.

249. Rigg CD, Cruickshank S. Carbon dioxide during and after the apnoea test—an illustration of the Haldane effect. *Anaesthesia.* 2001;56:377.

250. Goudreau JL, Wijdicks EF, Emery SF. Complications during apnea testing in the determination of brain death: predisposing factors. *Neurology.* 2000;55:1045–1048.

251. Brilli RJ, Bigos D. Apnea threshold and pediatric brain death. *Crit Care Med.* 2000;28:1257.

252. Goudreau JL, Wijdicks EF, Emery SF. Complications during apnea testing in the determination of brain death: predisposing factors. *Neurology.* 2000;55:1045–1048.

253. Zisfein J, Marks SJ. Tension pneumothorax and apnea tests. *Anesthesiology.* 1999;91:326.

254. Rudolf J, Haupt WF, Neveling M, Grond M. Potential pitfalls in apnea testing. *Acta Neurochir (Wien).* 1998;140:659–663.

255. Vardis R, Pollack MM. Increased apnea threshold in a pediatric patient with suspected brain death. *Crit Care Med.* 1998;26:1917–1919.

256. Bar-Joseph G, Bar-Lavie Y, Zonis Z. Tension pneumothorax during apnea testing for the determination of brain death. *Anesthesiology.* 1998;89:1250–1251.

257. Speelberg B, van Wezel HB. [Continuous pressure is preferred to flow triggering of respiration in the apnea test following the protocol for brain death determination]. *Ned Tijdschr Geneeskd.* 1998;142:1392–1393.

258. Telleria-Diaz A. [Apnea testing to establish death based on brain criteria]. *Rev Neurol.* 1998;27:108–110.

259. Kawamoto M, Sera A, Kaneko K, Yuge O, Ohtani M. Parasympathetic activity in brain death: effect of apnea on heart rate variability. *Acta Anaesthesiol Scand.* 1998;42:47–51.

260. Lang CJ. Blood pressure and heart rate changes during apnoea testing with or without CO_2 insufflation. *Intensive Care Med.* 1997;23:903–907.

261. Combes JC, Nicolas F, Lenfant F, Cros N, D'Athis P, Freysz M. [Hemodynamic changes induced by apnea test in patients with brain death]. *Ann Fr Anesth Reanim.* 1996;15:1173–1177.

262. Nicolas F, Combes JC, Louvier N, D'Athis P, Freysz M. Severe pulmonary arterial hypertension during the apnea test for brain death. *Transplant Proc.* 1996;28:375.

263. Wijdicks EF. In search of a safe apnea test in brain death: is the procedure really more dangerous than we think? *Arch Neurol.* 1995;52:338–339.

264. Lang CJ. Apnea testing by artificial CO_2 augmentation. *Neurology.* 1995;45:966–969.

265. Brilli RJ, Bigos D. Altered apnea threshold in a child with suspected brain death. *J Child Neurol.* 1995;10:245–246.

266. Maekawa T, Sadamitsu D, Tateishi A. [Safety applications in apnea test]. *Rinsho Shinkeigaku.* 1993;33:1331–1333.

267. Chantarojanasiri T, Preutthipan A. Apnea documentation for determination of brain death in Thai children. *J Med Assoc Thai.* 1993;76 Suppl 2:165–168.

268. Perel A. Apnea tests for brain death. *Arch Neurol.* 1991;48:1215–1216.

269. Keilson MJ. Apneic oxygenation in apnea tests. *Arch Neurol.* 1991;48:789–790.

270. Benzel EC, Gross CD, Hadden TA, Kesterson L, Landreneau MD. The apnea test for the determination of brain death. *J Neurosurg.* 1989;71:191–194.

271. Oikkonen M, Aittomaki J, Saari M. Apnoeic oxygenation of a brain-dead patient. *Lancet.* 1987;2:908–909.

272. Yokoyama T, Uemura K, Ryu H, et al. [Apnea test essential in the diagnosis of brain death]. *No To Shinkei.* 1987;39:959–963.

273. Belsh JM, Blatt R, Schiffman PL. Apnea testing in brain death. *Arch Intern Med.* 1986;146:2385–2388.

274. Outwater KM, Rockoff MA. Apnea testing to confirm brain death in children. *Crit Care Med.* 1984;12:357–358.

275. Rowland TW, Donnelly JH, Jackson AH. Apnea documentation for determination of brain death in children. *Pediatrics.* 1984;74:505–508.

276. Pitts LH, Kaktis J, Caronna J, Jennett S, Hoff JT. Brain death, apneic diffusion oxygenation, and organ transplantation. *J Trauma.* 1978;18:180–183.

277. Schafer JA, Caronna JJ. Duration of apnea needed to confirm brain death. *Neurology.* 1978;28:661–666.

278. Lang CJ, Heckmann JG. How should testing for apnea be performed in diagnosing brain death? *Adv Exp Med Biol.* 2004;550:169–174.

279. Lang CJ, Heckmann JG, Erbguth F, et al. Transcutaneous and intra-arterial blood gas monitoring—a comparison during apnoea testing for the determination of brain death. *Eur J Emerg Med.* 2002;9:51–56.

280. Lang CJ. Apnea testing guided by continuous transcutaneous monitoring of partial pressure of carbon dioxide. *Crit Care Med.* 1998;26:868–872.

281. Dominguez-Roldan JM, Barrera-Chacon JM, Murillo-Cabezas F, Santamaria-Mifsut JL, Rivera-Fernandez V. Clinical factors influencing the increment of blood carbon dioxide during the apnea test for the diagnosis of brain death. *Transplant Proc.* 1999;31:2599–2600.

282. Belsh JM, Schiffman PL. Apnoea testing to confirm brain death in clinical practice. *J Neurol Neurosurg Psychiatry.* 1987;50:649–650.

283. al Jumah M, McLean DR, al Rajeh S, Crow N. Bulk diffusion apnea test in the diagnosis of brain death. *Crit Care Med.* 1992;20:1564–1567.

284. Perel A, Berger M, Cotev S. The use of continuous flow of oxygen and PEEP during apnea in the diagnosis of brain death. *Intensive Care Med.* 1983;9:25–27.

285. Ropper AH, Kennedy SK, Russell L. Apnea testing in the diagnosis of brain death. Clinical and physiological observations. *J Neurosurg.* 1981;55:942–946.

286. Earnest MP, Beresford HR, McIntyre HB. Testing for apnea in suspected brain death: methods used by 129 clinicians. *Neurology.* 1986;36:542–544.

287. Shapiro BA. The apnea-PaCO₂ relationship: some clinical and medico-legal considerations. *J Clin Anesth.* 1989;1:323–327.

288. Coimbra CG. Implications of ischemic penumbra for the diagnosis of brain death. *Braz J Med Biol Res.* 1999;32:1479–1487.

289. Seshia S, Teitelbaum J, Young B. Response to Lessard, et al. *Can J Neurol Sci* 2000;27:353–354.

290. Brandstetter RD, Choi MY, Mallepalli VN, Klass S. An unusual cause of pulmonary barotrauma during apneic oxygenation testing. *Heart Lung.* 1994;23:88–89.

5
Ancillary Tests in Brain Death Confirmation

It is widely accepted that BD is a clinical diagnosis, and that confirmatory tests are not mandatory in most situations. According to Wijdicks,[1] "a confirmatory test is needed for patients in whom specific components of clinical testing cannot be reliably evaluated."[1] An ideal confirmatory study for BD should be safe, extremely accurate and reliable, available, quick and inexpensive.[2-9]

The therapeutic use of barbiturates in patients with severe intracranial hypertension or other forms of drug intoxication, hypothermia, and other metabolic disturbances, can prevent determination of brain death by clinical criteria.[10-21] In certain European, Central and South American, and Asian countries, the law requires confirmatory tests.[22,23] The diagnosis of BD in children and neonates is more complicated and ancillary tests are usually suggested.[6,24-29]

Confirmatory tests in BD can be divided into those that prove absent cerebral blood flow (CBF) and those that demonstrate loss of bioelectrical activity. In Cuba, we proposed using confirmatory tests (which are still optional) when the clinical examination is not reliable, to shorten the period of observation, and in primary brainstem lesions.[30-33]

In fact, confirmatory tests that are widely accepted are conventional angiography and electroencephalogram (EEG).[1,4,5,34] Nonetheless, we have defended combining transcranial Doppler (TCD) ultrasonography and a neurophysiological test battery (multimodality evoked potentials and electroretinography) to increase BD diagnostic reliability.[30,31,35-40]

Tests to Demonstrate a Complete Cessation of Brain Circulation

During the 1950s and 1960s the phenomenon of cerebral circulatory arrest (or blocked cerebral circulation) was repeatedly demonstrated.[41] In 1956, Löfstedt and von Reis[42] in Sweden, described six patients who demonstrated no passage of contrast to the cerebral circulation. The authors stated that there was a slow circulation of blood that was not possible to demonstrate at angiography, due to an increased intracranial pressure (ICP), possibly in combination with vasospasm. They did not diagnose their patients as dead.[42–44] Wertheimer et al.,[45] after describing the death of the nervous system in 1959, published a case report in 1960 illustrating how the ventilator had been turned off in a 13-year-old patient. To confirm the diagnosis of death of the brain, they performed angiography in both the internal carotids and the vertebral arteries, there was no circulation of blood. In the 1970s, contrast arteriography was commonly used to demonstrate the absence of intracranial circulation in the BD diagnosis.[46–59] Nowadays, this technique continues to be used in some hospitals lacking access to simpler alternative techniques to study brain circulation.[23,60–65]

Bernat[66] recently emphasized that "the most confident way to demonstrate that the global loss of clinical brain functions is irreversible is to show the complete absence of intracranial blood flow." It is well established that brain neurons are irreversibly damaged after a few minutes of complete cessation of CBF, and are globally destroyed when blood flow completely ceases for about 20 to 30 minutes.[66–69] Ingvar[58,70] found that the permanent cessation of CBF produces a total brain infarction. Other authors have reported that cardiac arrest with normal body temperature, lasting about 10 to 15 minutes, produces total cerebral ischemia, leading to an irreversible total loss of brain function. Hence, if CBF has stopped for more than 30 minutes, this proves the irreversibility in the BD diagnosis.[66,68,71–73] The rationale is that absence of the cerebral circulation for this period of time causes a total brain infarct. The cerebral circulatory arrest clearly had to be total (i.e., to affect both the carotid and vertebral circulations) in order to document the "death of all intracranial neurons."[66,74,75]

Several mechanisms have been proposed to explain the arrest of brain circulation in BD. The first is a massive increase in ICP, leading to increased cerebral vascular resistance, the extreme slowing of flow, and arteriolar and capillary perfusion pressure, provoking the cessation of cerebral circulation. When ICP is lower than systolic blood pressure but higher than diastolic pressure and exceeding the mean arterial pressure, blood comes into the cranium and brain during systole but is driven back an identical amount during diastole. This phenomenon is known as "reverberating flow," because there is no perfusion into the brain, that is, there is no net forward circulation. Thus, because there is no net blood flow, the brain is diffusely and irreversibly destroyed within minutes.

The second mechanism is a progressive intravascular obstruction that accompanies the death of the brain and leads to an increase in cerebral vascular resistance, due to the effect of progressive ischemia, producing tissue swelling and subintimal bleb formation. The third mechanism is the destruction of the brain vessels themselves due to the whole brain infarct. The pathological end result is the so-called respirator brain, characterized by a soft and necrotic organ that autolyzes at body temperature when patients are kept under ventilatory assistance for many hours or days after brain circulation has ceased.[7,48,50,57,58,66,67,70,76]

Several tests have been developed in the last few decades that can accurately and validly measure CBF in suspected brain-dead patients. The first technique used to demonstrate absence of intracranial circulation in BD distal to the intracranial portions of the internal carotid and vertebral arteries was the cerebral angiography (CA).[42–44]

Cerebral Angiography

Four-vessel angiography has been considered the most reliable investigation in the diagnosis of BD for more than three decades. This technique facilitates showing the absence of intracranial circulation distal to the intracranial portions of the internal carotid and vertebral arteries.[60]

Several methods have been used for introducing contrast media in the carotid and vertebrobasilar systems: selective catheterization of all four arteries by femoral or axillary artery puncture, direct puncture of the carotid and the vertebral arteries, or cerebral panarteriography by injection of contrast media into the ascending aorta or by retrograde brachial injection.[4,60–63,77–80] It has been noted that the Seldinger technique through the femoral arteries has been the most frequently used method, with a catheter passed through the femoral artery into the ascending aorta, into which dye is injected under pressure.[48,81–83] Moreover, the field in which the arterial puncture is carried out is well away from the head, where resuscitative procedures are in progress. Several authors have also emphasized that puncture and intubation of the femoral artery are relatively simple, and, monitoring with x-ray control, selective catheterization of the arteries at the neck can be achieved easily. The dye is usually injected under two or three atmospheres of pressure, and, because of the slowed circulation, serial films should be taken during longer periods of time when CA is performed for other diagnostic procedures. In addition, when injecting the dye into the aorta, the opacification of the external carotid circulation serves as a control.[5,48,84] Delayed filling of the superior longitudinal sinus has been occasionally found.[85–87]

The most commonly pattern for the carotid system is an abrupt block of circulation at the cranial base or in the carotid siphon near the anterior or posterior clinoid process with or without visualization of the ophthalmic artery. Usually the dye column arrests symmetrically in the carotid arteries, but occasionally it extends more distally on one side. In some cases, a narrowing of the dye column and nonfilling, considered to be a "pseudo-occlusion"[7,47] in the cervical portion of the internal carotid artery,[88] are demonstrated. Walker[7,47] commented that in a few cases a delayed filling of the intracranial portion of the internal carotid artery is shown with a slight density in the first part of the middle cerebral artery and, rarely, opacification of the anterior cerebral artery. As has already been emphasized, excellent visualization of the external carotid vessels serves as a control to demonstrate the absence of intracranial circulation.[5] In brain-dead patients no intracranial venous phase is ever observed.[7]

Regarding the vertebrobasilar system, the dye column in the vertebral artery usually is interrupted at the atlanto-occipital junction, and less often intracranially; it is rarely blocked in the transversarial canal. In a few cases, a narrow column of dye is shown in the basilar artery lying against the clivus, and the vertebral artery is opacified by retrograde filling. Intracranial circulation of the carotid and basilar systems may stop at the same time that the carotid flow has ceased, but the vertebrobasilar circulation persists for a few hours more. In the vertebrobasilar system the difference between extra- and intracranial circulation is less prominent, because if intracerebral vascular filling takes place, the dye opacifies the vessels for up to 30 minutes.[7,53,60,63,79,89,90]

In some cases fulfilling clinical and EEG criteria of BD carotid angiography occasionally revealed a blocked circulation while vertebral injections still revealed some flow.[55,91–100] Rarely, the converse occurred, that is to say, carotid circulation without vertebral flow.[100] It has been emphasized that pressures are not essentially equivalent in the supratentorial and infratentorial compartments. In those case, when applying whole BD criteria, the diagnosis cannot be definitively established.[81]

Hence, although many physicians have stated that CA is the critical method of diagnosing absent brain circulation in BD, its main disadvantage is its invasiveness and that the patient must be transported from the intensive care unit (ICU) to the x-ray department.[66] Possible complications include hypotension, bradycardia, the harmful effects of introducing hypertonic dye material into partially obstructed vessels, vasospasm, subintimal injection, and thrombosis. Hence, the use of contrast angiography seems dubious from the ethical point of view, because as a confirmatory test in BD diagnosis, when the patient is not yet dead, the patient is exposed to possible risks with no prospective benefits.[4,61,62,66,77,80,101–103]

Cerebral Intravenous Digital Subtraction Angiography

Several authors have suggested using cerebral intravenous digital subtraction angiography (IV DSA) in BD confirmation, instead of conventional cerebral angiography.[104,105] Ducrocq et al.[106] proposed injecting 60 to 80 mL of contrast medium cannulating a brachial vein, using an automatic syringe injector at a flow rate of 12 to 15 mL/s. In the meantime simultaneous anteroposterior and lateral or lateral projections alone are performed, achieving a total exposure delay of about 6 to 8 seconds (each run lasts 60 seconds). These authors emphasized that compared with IV DSA, cerebral angiography is a more invasive, time-consuming, and expensive test than IV DSA, and also requires a greater expertise.

The IV DSA findings in BD consist of visualizing carotid and vertebral blood flow interruption at the cervical, skull base, or intracranial levels. This method demonstrates vascular stasis, characterized by no progression of the contrast medium during a prolonged series of 1-minute duration. Some authors have found a slow posterior fossa circulation in patients fulfilling all clinical signs of BD.[107] Bunnen et al. studied 110 patients with clinical signs of brain death, and in 105 patients this technique demonstrated intracranial circulation arrest; in the other five patients, repeated examination a few hours later confirmed the cessation of brain circulation.[108]

Because this is not an invasive test, it has been proposed as a BD confirmatory test in children, taking into consideration that a BD diagnosis at this age is especially elusive and difficult.[104] Some authors have found portable DSA equipment very useful in ICUs to demonstrate the absence of brain circulation during BD confirmation.[105]

Several authors have defended this technique because it requires only an injection using a catheter placed in a peripheral vein, and it is a fast, safe and easy method to apply in BD confirmation.[77,104–106,108–115]

Intravenous Radionuclide Angiography

Goodman, Mishkin, and Dyken-Plum studied cerebral perfusion by isotope radionuclide angiography. The technique consists of injecting technetium-99m (^{99m}Tc) pertechnetate as a bolus in the antecubital vein and measuring isotopic emissions from the head, using a gamma camera, every 3 seconds for 24 seconds.[103,116–119] Korein's group[120] compared the result with conventional arteriograms and concluded that this bolus technique is a helpful confirmatory test in diagnosing BD. They stated that the failure to demonstrate intracranial radioisotopic activity on isotope angiography is thought to imply a critical reduction of CBF to less than 24% of normal perfusion.

This test is possible to carry out even if a patient is experiencing drug intoxication, hypothermia, or severe metabolic disorders, extending its diagnostic possibilities. Several authors have defended its use for the diagnosis of BD in children.[121,122]

Wijdicks[5] stated that one of the main difficulties using this technique for diagnosing BD is evaluating the implication of grossly reduced, but not totally absent, cerebral blood flow (CBF). When there is low flow oxygen consumption ($CMRO_2$), it is the most significant index of the brain metabolic activity and it is reliant on CBF. Hence, he concluded that grossly reduced head bolus patterns characterized by traces of radioactivity remaining in the venous sinuses do not exclude the diagnosis of BD. Other drawbacks of this technique are that it does not represent in detail cerebral vessels anatomy, and it is not suitable in assessing the perfusion of structures in the posterior fossa.[5,6]

Single Photon Emission Computed Tomography

Single photon emission computed tomography (SPECT) provides multiplanar imaging of brain tissue perfusion and has been proposed as a good candidate for the gold standard as a confirmatory test in BD. Another main advantage of this technique is that it is noninvasive, and it can assess both supra- and infratentorial vascular areas.[60,123]

Using blood flow tracers, loss of perfusion can be shown with high regional sensitivity and specificity. The technique consists of injecting intravenously a tracer isotope, ^{99m}Tc–hexamethylpropyleneamine oxime (HMPAO), at least 15 to 30 minutes before scanning. Serial pictures are taken dynamically, followed by anterior and lateral static images, creating a three-dimensional image.[25,29,60,77,123-130] Mean arterial blood pressure during injection must be above 80 mm Hg in adults and 60 mm Hg in children in order to prevent temporary cerebral hypoperfusion.[131] Since extracranial perfusion is preserved, there are residual counts that might be misinterpreted as brain uptake of ^{99m}Tc-HMPAO; other SPECT tracers sustain high first-pass extraction inside the brain due to their lipophilicity. However, tracer retention in the brain is produced by a change of the molecules to a hydrophilic form, which is done by functioning cell energy metabolism.[132,133] Hence, some authors have argued that in BD, energy metabolism has stopped in the dying or dead neurons, and that loss of tracer uptake is due to cell death, and not because of absent perfusion.[131] Nonetheless, Weckesser and Schober[134] have emphasized that the loss of tracer uptake inside the brain immediately after tracer injection in early static images, when the role of tracer washout is not yet important, is explained by absent brain perfusion.[134]

In BD, SPECT demonstrates the arrest of brain circulation ("empty skull") in about 96% of cases; in some suspected brain-dead patients, perfusion may persist in the thalamus and brainstem, particularly in infants. This finding in BD diagnosis will only lead to postponing a definitive diagnosis. However, this is not the kind of error feared by physicians, nurses, and patients' relatives.[123]

Okuyaz et al.[25] recently found that this technique has high reliability in BD confirmation in infants, although they propose a second study to confirm the diagnosis in newborns. Other authors have recently concluded that although SPECT provides consistency as a confirmatory test in BD diagnosis, further studies including larger series of cases should be done to verify its reliability in accessing the arrest of brain circulation.[60]

A drawback of this technique is that although mobile gamma cameras are commercially available, BD confirmation commonly requires that the patient must be transported from the ICU to the SPECT lab, because this modality is not widely available in emergency services.[5]

Transcranial Doppler Ultrasonography

Transcranial Doppler (TCD) ultrasonography is a noninvasive technique that measures local blood flow velocity and direction in the proximal portions of large intracranial arteries. It requires training and experience to perform it and to interpret results; hence, it is typified as operator-dependent.[135-137] In the ICU setting, intensivists or neurologists usually receive training in applying this technique using portable Doppler devices in suspected brain-dead cases.[4,136-141] In general, the principal advantages of TCD are that it is noninvasive, it can be carried out at the bedside, it can be repeated as needed or in continuous monitoring; it is less expensive than other techniques; and dye contrast agents are not needed. Its main disadvantage is that it can study CBF velocities only in certain segments of large intracranial vessels.[135-137]

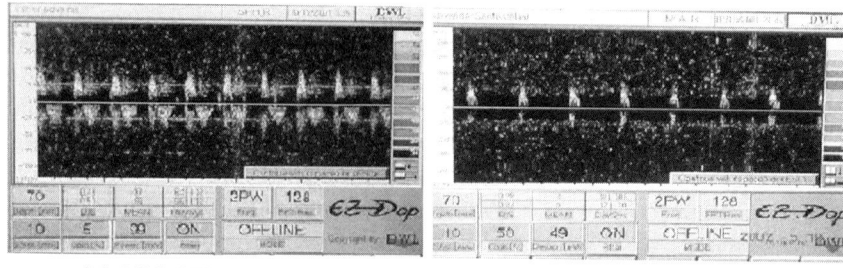

OSCILLATING FLOW   **SYSTOLIC SPIKES**

FIGURE 5.1. Doppler-sonographic patterns in two brain-dead patients: Left (Oscillating flaw), Right (Systolic spikes). Courtesy of Dr. Anselmo Abdo. (See also color insert.)

Immediately after Doppler-sonography had been introduced in clinical practice, typical findings for brain circulatory arrest were described.[62,142–144] The American Academy of Neurology Therapeutics and Technology Assessment Subcommittee presented a remarkable report on TCD ultrasonography clinical applications. The use of TCD to diagnose cerebral circulatory arrest and BD was fully analyzed. The subcommittee reviewed a number of high-quality reports that also discuss some caveats that have an important impact on the diagnosis of BD by TCD, concluding that with strict criteria, TCD is highly sensitive and specific for the diagnosis of BD.[137]

As a confirmatory test, TCD in BD demonstrates evidence of absent CBF (zero net flow velocity) at all insonation sites. Operators must explore insonation of both middle cerebral arteries through the temporal bone window, above the zygomatic arch.[137,145–150]

Oscillating flow and systolic spikes patterns are typical Doppler-sonographic flow signals found in the presence of cerebral circulatory arrest, which, if irreversible, results in brain death (Fig. 5.1). The pathophysiology to explain these findings is as follows: In comatose patients, the earliest sign of an ICP augmentation is an increased pulsatility followed by progressive decrease in diastolic flow velocities and reduction in mean flow velocities; pulsatility changes occur when cerebral perfusion pressure is 70 mm Hg. If the velocity at the end of diastole becomes zero, then the ICP has reached the diastolic blood pressure. Forward flow continues in systole, and hence, in this phase a brain circulatory arrest cannot be diagnosed. When the ICP equals or exceeds the systolic blood pressure and forward and reverse flow are nearly identical, then in this stage a cessation of cerebral perfusion has been reached. It is characterized by a pattern known as oscillating flow, biphasic flow, or net zero flow. Equality of forward and reverse flows can be demonstrated, calculating the area under the envelope of the positive and negative deflection in the velocity waveforms. As an additional reduction of the blood movement occurs, systolic spikes appear, which are very short velocity peaks. The systolic spike is a definitive pattern for diagnosing brain circulatory arrest. Finally, when ICP augments further and flow hitch becomes more proximal,

TABLE 5.1. American Electroencephalographic Society guidelines for recording EEG in suspected brain-dead patients[200]

1. A minimum of eight scalp electrodes and ear reference electrodes
2. Interelectrode resistances under 10,000 ohms but over 100 ohms
3. Test for integrity of the recording system by deliberate manipulative creation of electrode artifact
4. Interelectrode distances of at least 10 cm
5. Gains increased during most of the recording from 7.0 μV to 2.0 μV/mm
6. The use of 0.3– or 0.4–sec time constants during part of the recording
7. Recording of an ECG and of extracerebral potentials by a pair of electrodes on the dorsum of the right hand
8. Tests for reactivity to pain, loud noises, and light
9. A 30-minute total recording time
10. Recording by a qualified technician
11. Repeating the record if electrocerebral silence (ECS) is doubtful
12. Telephone transmitted EEGs are not appropriate for determination of ECS.

no flow signals in the basal cerebral arteries are identified. It is important to stress that a failure to detect flow signals can be due to ultrasonic transmission problems. To face this controversy and confirm the diagnosis, it is necessary to perform extracranial bilateral recording of the common carotid, internal carotid, and vertebral arteries.[62,64,135,137,147,149–161]

The Neurosonology Research Group of the World Federation of Neurology created a task force to evaluate the role of Doppler sonography as a confirmatory test for determining brain death, concluding that "extra- and intracranial Doppler sonography is a useful confirmatory test to establish irreversibility of cerebral circulatory arrest as optional part of a brain death protocol." Moreover, this task force especially recommended TCD in patients when the therapeutic use of sedative drugs causes the EEG to be unreliable. This group proposed a series of guidelines for the use of Doppler-sonography for detecting brain circulatory arrest (Table 5.1).[162]

The consensus statement by the task force on cerebral death of the Neurosonology Research Group of the World Federation of Neurology includes the following:

• Cerebral circulatory arrest can be confirmed if the extra- and intracranial Doppler sonographic findings have been recorded and documented both intra- and extracranially and bilaterally on two examinations at an interval of at least 30 minutes.
• Systolic spikes or oscillating flow in any cerebral artery can be recorded by bilateral transcranial insonation of the internal carotid artery (ICA) and middle cerebral artery (MCA), or any branch or other artery that can be recorded (anterior and posterior circulation).
• The diagnosis established by the intracranial examination must be confirmed by the extracranial bilateral recording of the common carotid, internal carotid, and vertebral arteries.

- The lack of a signal during transcranial insonation of the basal cerebral arteries is not a reliable finding because it can be due to transmission problems. But the disappearance of intracranial flow signals in conjunction with typical extracranial signals can be accepted as proof of circulatory arrest.
- Ventricular drains or large openings of the skull, as in decompressive craniectomy possibly interfering with the development of the ICP, are not present.

Other Techniques to Detect Absent Brain Circulation

Indirect methods to measure CBF such as echoencephalography, measurement of arm to retina circulation time, ophthalmic artery pressure, and rheoencephalography have not been shown to be consistent enough to be suitable in BD confirmation.[5–7,47,53,163–164] An earlier technique used in BD patients was xenon-enhanced computed tomography (CT).[165–171]

Since the introduction of CT scan in medical practice, this technique has been used to document structural, irreparable brain damage, in comatose patients, as a prerequisite in BD diagnostic criteria standards.[5,60,171–176] Standard magnetic resonance imaging (MRI) has been used for the same purpose, demonstrating diffuse swelling of the cerebral gyri and cerebellar cortex, with prolongation of both the T1 and T2 signal (representing hypoxic ischemic brain injury), downward displacement of the diencephalon and the brainstem (central and tonsillar herniation), and loss of blood flow in the intracranial portions of both internal carotid arteries.[5,6,177–182]

Nonetheless, several CT- and MRI-related techniques have been developed to assess brain circulation in BD confirmation.[23,77–79,175,183] Several authors have proposed using contrast-enhanced CT scans to demonstrate the lack of cerebral artery perfusion.[175,187–189] Qureshi et al.[23] have emphasized that CT angiography with CT perfusion may represent a rapid noninvasive method for BD diagnosis.

In BD patients magnetic resonance (MR) arteriography showed no evidence of arterial flow in the intracranial circulation above the level of the supraclinoid portion of the ICAS and in the vertebrobasilar arteries.[178] These authors reported three patients who underwent MR venography who presented partial superior sagittal sinus opacification. The MR angiograms did not show the intracranial vessels above the level of the supraclinoid portion of the ICAs.

Watanabe et al.,[190] applying three-dimensional anisotropy contrast MR axonography in BD patients, concluded that diffusion anisotropy of neural tracks lessened between 1 and 12 hours the BD onset, probably related to the bundle length, and disappeared about 24 to 44 hours after BD commencement, which might be a sign of dynamic changes of axonal structure, inducing axonal dysfunction.[190]

Kumada et al.,[191] studying BD patients by diffusion-weighted images (DWI) and apparent diffusion coefficient (ADC), found that after BD onset DWI showed high signals in the cerebral cortex, putamen, thalamus, brainstem, and cerebellum, and decreased ADC values up to 30% to 40%. They emphasized that DWI and ADC mapping shows areas corresponding to edema of a cytotoxic nature and to ischemic tissue. Nakahara et al.[192] also reported a BD case with ADC values of gray and white matter that were significantly lower than those of three other brain-injured patients who were not brain dead; additionally, the ADC value of white matter was significantly lower than that of gray matter. Other investigators have reported similar results.[190–196]

Confirmatory Tests to Demonstrate Loss of Bioelectrical Activity

Electroencephalogram

The EEG has been closely linked to BD since the pioneering descriptions of the death of the nervous system and coma *dépassé*.[45,197] In 1963, Robert S. Schwab et al.,[198] proposed that "the total absence of EEG activity after 30 min of recording is most important evidence of death of the nervous system." They went further than to recommend EEG as an ancillary test: they incorporated EEG findings into their diagnostic criteria, which were similar to those of the Harvard Committee (Fig. 5.2).

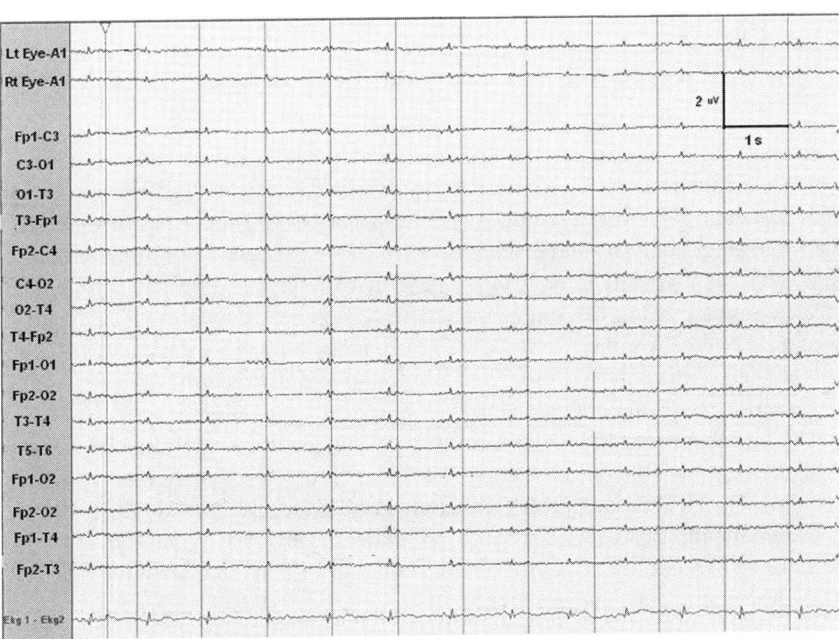

FIGURE 5.2. Electrocerebral silence in a brain-dead patient using digital electroencephalogram (EEG). See electrocardiogram (ECG) contamination in all leads. (See also color insert.)

In fact, the history of BD acceptance has emphasized in many cases the need for an objective confirmation of BD arising from the need to reassure patients' relatives and others about the certainty of diagnosis, as well as to protect the attending doctor from prosecution (essentially for murder).[59,176,199-210] Moreover, the use of EEG is closely related to the conceptual basis for accepting death on neurological grounds. A set of criteria based on a brainstem standard, such as those from the Commonwealth countries, does not include EEG.[85,211-216] In contrast, EEG is recommended by all countries adopting the "whole brain death" definition.[217-222] For example, EEG is mandatory in Italy.[223]

The Harvard Committee included EEG as a technical investigation able to explore brain function and to give an objective confirmation of BD.[224] However, in 1969, Beecher,[225] the chairman of the ad hoc committee, asserted that the EEG, although valuable, was not essential. Later, several authors[107,215,226-229] emphasized the technical pitfalls and limitations of the EEG, which can be summarized as follows:

- The EEG does not provide direct information on brainstem function.
- The EEG may be flat in patients with preserved brainstem function (e.g., comatose or vegetative patients following prolonged cardiac arrest).
- Residual EEG activity can be found in primary brainstem lesions fulfilling most clinical BD criteria.
- The EEG is unreliable in cases of sedation, hypothermia, or presence of toxic or metabolic factors.
- There is a consistent risk of mistaking artifacts for residual cortical activity and vice versa.

In 1969 the American Electroencephalographic Society carried out a huge survey to evaluate the use of EEG in BD confirmation.[208] They included in the survey 1665 patients with EEG records pronounced flat by 100 reporting members of the society, and only seven persons recovered with the following coma etiology: three with sedative drug intoxication, one with insulin coma, one who suffered asphyxia from drowning, one after cardiac arrest, and one who was in shock during a surgery. Nonetheless, the society agreed to consider only those three patients who complained of drug intoxication, because the rest of the records were not useful to diagnose an electrocerebral silence (ECS).

The Collaborative Study, based on a review panel of experts, estimated that the disagreement with the primary reader would have been less than 3%.[9] Chatrian[230] concluded that if ECS is clearly demonstrated, subsequent recovery of EEG activity is extremely improbable in individuals not under the effect of hypothermia, drugs, or extreme hypotension. Buchner and Schuchardt[226] found that sensitivity as well as the specificity can reach up to 90%.

The EEG tracing related to BD is a straight line, sometimes contaminated with different artifacts, that has been given various names, such as flat, isoelectric, null, complete inactivity, electrocerebral inactivity, electrocerebral silence, electrical silence, without detectable cortical activity, and absence of biological activity.[7] Nonetheless, the American Electroencephalographic Society's

Committee on Cerebral Death recommended using the term *electrocerebral silence,* defined as "no electrocerebral activity over two microvolts (μV) when recording from scalp or referential electrode pairs 10 or more centimeters apart with interelectrode resistances under 10,000 ohms (or impedances under 6,000 ohms) but over 100 ohms."[200] Technically, this definition of ECS does not mean that the brain is not generating any bioelectrical activity, but only that any EEG activity with less than 2 μV cannot be distinguished from the recording instrument noise. In routine EEG labs the amplification of 7.0 μV/mm is so low that signals of 5 μV produce pen excursions of less than 1 mm, which are very difficult to detect as reliable bioelectrical activity.[7,200,206–209,215]

Due to this technical complexity in recording a reliable ECS, several authors have proposed specific guidelines to record EEG in suspected BD patients. Trojaborg[203,231,232] pointed out the importance of recording at high sensitivity, from more scalp areas, for a sufficient length of time, and to take into consideration extracerebral potentials that are frequently present in those cases.

In 1976 the American EEG Society formulated specific guidelines for recording EEG in suspected brain-dead patients[200] (Table 5.1).

Hence, EEG readers must be aware that in the intensive care environment artifacts are common in EEG records. When reversible causes of coma are excluded, the presence of artifacts, very common in the intensive care environment, may account for a nonnegligible uncertainty about the presence or absence of electrocerebral activity.[223] Artifacts may come from the electroencephalograph, the connectors to the patient, the environment, and the patient. Very common artifacts are due to high electrode-skin impedance, electrode movements due to nursing procedures, or feedback or inductance from equipment for monitoring the vital signs. Another frequent artifact that might resemble real EEG activity is regular and rhythmic waves time-locked with the respirator cycle; it may be removed by stopping the ventilator for a short time. The electrocardiogram (ECG) is another frequent source of artifacts, because potentials generated by the patient's heartbeat are usually greater in BD patients probably due to changes in intracranial impedance resulting from an absence of CBF.[5,7,215]

Digital EEG and Neuromonitoring

Serial routine EEGs had been proposed to follow up comatose patients for an early detection of BD onset. Nonetheless, continuous visual monitoring of EEG tracings on conventional paper is an impractical task in the intensive care environment. Hence, several EEG monitoring techniques have been developed to be applied in ICUs.[233–243]

With the development of digital EEG and several EEG monitoring techniques in the intensive care environment, such as compressed spectral analysis (CSA), bispectral index (Bis), and cerebral function monitor, several authors have reported that these procedures can be used in severely comatose patients as an assessment of BD onset, enabling appropriate scheduling of either EEG or other confirmatory tests to confirm BD.[243–255]

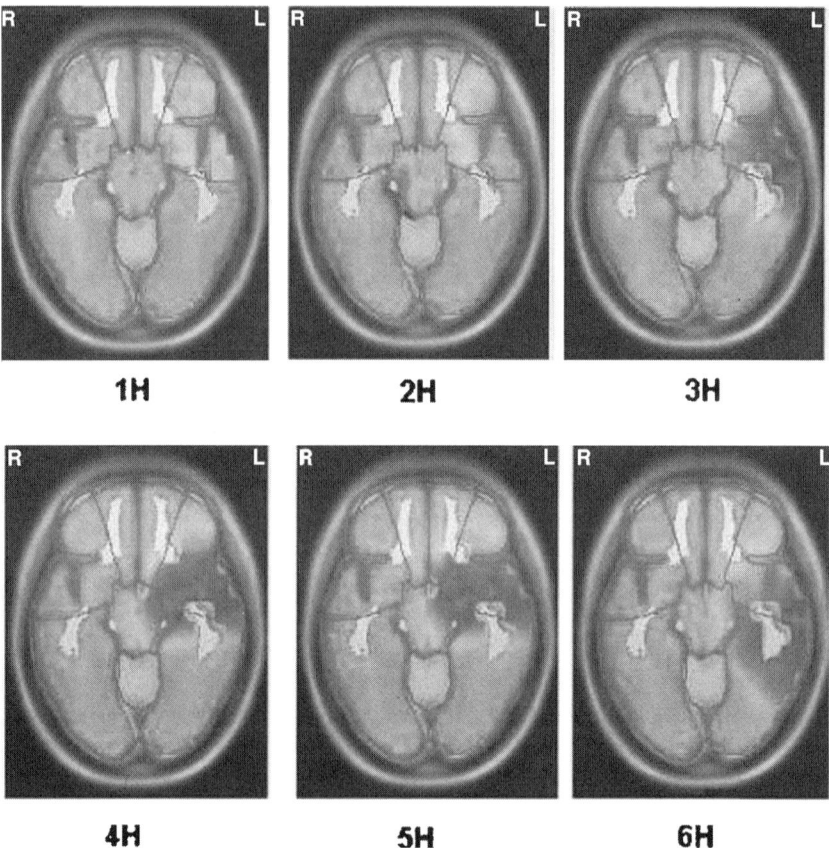

FIGURE 5.3. Continuous monitoring in a comatose patient. At 6 hours, an electrocerebral silence (see Fig. 5.2) is recorded.[256] (See also color insert.)

The term *continuous EEG monitoring* (cEEG) has been recently introduced in the literature. As cEEG is noninvasive, inexpensive, and continuous, it is the natural choice for routine brain monitoring in comatose patients. However, cEEG has remained largely unused for this purpose outside of the operating room. Currently, the main obstacles to cEEG in the ICU are logistical and to a lesser extent technological: maintaining working electrodes on the head, having EEG experts available 24 hours a day, and developing effective detection algorithms to set off bedside alarms. Nonetheless, cEEG is the best choice to monitor brain electrical activity in comatose patients developing BD. It is, without doubt, the best technical tool for early detection of a loss of EEG activity (Fig. 5.3).[233,234,237,238,256]

It is now appropriate to redefine the role of EEG, taking into account the substantial progress in the definition of BD and the development of monitoring techniques, in order to eliminate all the sources of over- and underestimation of their effectiveness. Hence, EEG remains an important confirmatory test in BD diagnosis.

Multimodality Evoked Potentials and Electroretinography

Clinical examination is hampered by several factors, such as drug intoxication, barbiturate narcosis, and hypothermia.[35,37,38,40,257−262] Most sets of criteria require some time period that patients must show signs of BD (6, 12, 24 hours, etc.) in order to establish a definitive diagnosis.[263−279]

On the other hand, multimodality evoked potentials (MEPs) and electroretinography (ERG) are highly resistant to the above-mentioned factors and have been shown to be reliable in the intensive care unit environment.[35−40,223,257,258,280−284] Due to these features, substantial interest has grown over the past two decades in MEPs in BD, and a wealth of data is now available in the literature.[35,61,223,280,285−293] Nonetheless, considered as a single test, MEP and ERG have their limitations and they are not routinely included as confirmatory tests for BD diagnosis.[4−7,9,22] Hence, I propose combining MEP and ERG in a test battery to study BD patients in order to increase their diagnostic reliability.

The use of each modality, singly and in combination, in the diagnosis of BD will now be briefly reviewed.

Visual Evoked Potentials and Electroretinography

Visual evoked potentials (VEPs) may be a sensitive indicator of early impairment of cerebral function and may demonstrate changes sooner than an EEG, and therefore several articles have appeared describing the use of VEP to study BD patients.[35,36,40,203,231,283,292,294−297] Nevertheless, there are still controversies among investigators about the possible contamination of VEP recordings by spread of the ERG to the occipital area, which could limit the use of this technique for the diagnosis of BD.[35,40,203,231,298−300]

We found a characteristic pattern in all patients. When a cephalic reference was used for both VEPs and the ERG, the a and b waves of the ERG were recognized in all cases. The visual evoked responses consisted of waves with less amplitude but the same latency and morphologic features as in the ERG. When a noncephalic derivation was chosen for the ERG and VEPs, the ERG waves were the same in latency and morphologic characteristics, but the VEP channel showed no response (Fig. 5.4).[35,40,298−300]

Jørgensen and Trojaborg[231] noted that when the summated ERG was superimposed with reverse polarity on the response recorded over the occipital region, it was canceled, and they concluded that the VEP waves recorded in BD patients are the spread of the ERG to the occipital area. These investigators were in disagreement with others[301−306] who did not observe the spread of the ERG to the occipital region. Jørgensen and Trojaborg explained this discrepancy by the fact that they applied up to 500 to 1000 stimuli and used amplification at least 10 times greater than those of other studies. At our center, we found that the amplitude of the waves recorded in the VEP channel decreased progressively when the reference electrode was more distant from the retina. Moreover, in our patients the amplitudes of the ERG waves were greater in the ERG derivation with noncephalic

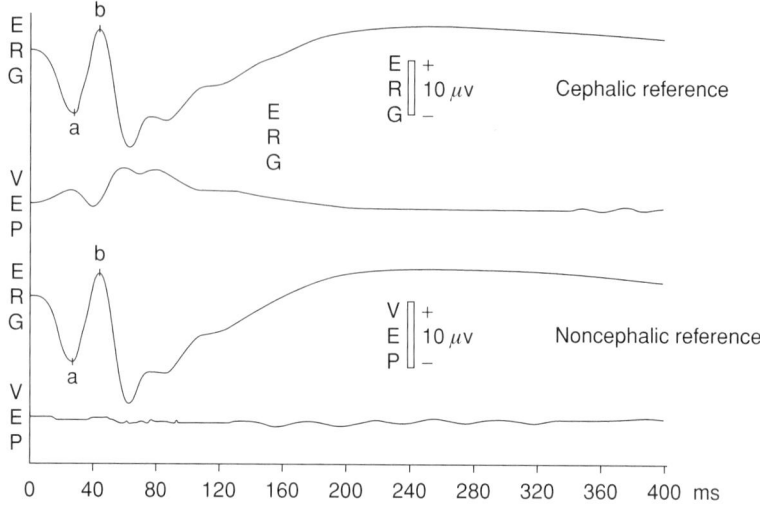

FIGURE 5.4. We found a characteristic pattern in all patients. When a cephalic reference was used for both visual evoked potentials (VEPs) and the electroretinography (ERG), the a and b waves of the ERG were recognized in all cases. The visual evoked responses consisted of waves with less amplitude but the same latency and morphologic features as in the ERG. When a noncephalic derivation was chosen for the ERG and VEPs, the ERG waves were the same in latency and morphologic characteristics, but the VEP channel showed no response. (Modified from reference # 321.)

reference, suggesting that the ERG was also picked up by the Fz electrode in the ERG channel with cephalic reference. Only in two cases was it possible to record waves in the VEP derivation with a noncephalic reference, indicating a spread of the ERG to the occipital area.[35,36,40,298–300]

Jørgensen and Trojaborg used A_1 or A_2 as references, and it is likely that the ERG was picked up mainly by the earlobe electrode and not by the occipital one. Because we used a noncephalic-referenced VEP derivation, the reference electrode was farther from the retina than the recording electrode placed in the occipital area. These results suggest that, although contamination of the VEP records by the spread of the ERG to the occipital area could occur, it is easy to confirm the absence of a true cortical visual response in BD patients by means of a noncephalic reference.[35,36,40,298–300]

On the other hand, persistence of the ERG, even long after the elimination of all respiratory support, has led some investigators to assert that this technique is not useful for studying BD patients.[307,308] The ERG, an electrical response arising from photic stimulation of the outer retinal layers, is dependent on an adequate retinal blood supply from the ophthalmic artery. Because the ophthalmic circulation in 3% of cases is entirely derived from the external carotid artery, and in the remainder, receives a collateral circulation from the facial artery, retinal ischemia may not develop until the systemic circulation fails. Hence, it is not surprising that the ERG has been found to give inconsistent results in BD cases.[309–317] Arfel[301] in

1967 reported that from patients in coma *dépassé*, an averaged biphasic response could be obtained from frontal scalp electrodes for as long as several days after the EEG was isoelectric. Sims et al.,[318,319] using computer techniques to produce a three-dimensional display (compressed spectral analysis) of the ERG, found that a BD patient can have an ERG indistinguishable from that of an alert individual.

We elicited and recorded VEPs and the ERG simultaneously, using a cephalic (Fz) and a noncephalic (C7) reference. In all cases, when a noncephalic reference was used, the ERG waves did not differ in either latency or morphologic features, while the VEP channel showed no response. This pattern clearly confirms that in the visual pathways of BD patients, electrical activity is confined to the retina. Our results further suggest the usefulness of combining VEPs and ERG in the same setting, using cephalic and noncephalic references to examine patients suspected to be brain dead.[35,36,40,298–300] In other modalities of evoked potentials, the use of noncephalic references has also been suggested for a better elucidation of evoked potential component generation.[320–324] Moreover, as for median nerve somatosensory evoked potentials (SEPs), in which the ERG's potential serves as a stimulation input control,[322] by recording the ERG it is possible to ensure that adequate visual stimuli have been applied.[35,36,40,298–300]

Although it is now more common to elicit the ERG with more sophisticated forms of stimuli, such as pattern reversal and color- and intensity-controlled light flashes,[325–327] light-emitting diodes seem to be suitable to study unresponsive patients in ICUs, where the conditions of illumination are difficult to manage. Moreover, goggles are easily attached to intubated and unconscious patients. When the visual pathways of suspected BD patients are studied by VEP and ERG recordings, it is possible to assess mainly forebrain hemispheric function, while brainstem evaluation is lacking. Therefore, it may be concluded that, although the setting we propose for VEPs and the ERG provides an objective electrophysiologic assessment of hemispheric function, which is essential for diagnosis of brain death, this technique is probably of limited value for this purpose when used in isolation.[35,36,40,298–300]

Brainstem Auditory Evoked Potentials

After Starr's[328] first clinical report on the use of brainstem auditory evoked potentials (BAEPs) in the determination of BD, several papers on this topic have appeared in recent years.[329–335] Some of these authors have found similar patterns of BAEP in their series of BD patients, that is, no detectable response (the most frequent) or only wave I, unilateral or bilateral, without any later components.[289,328,330] However, other investigators have also reported the presence of wave II.[335,336] We have reported the following BAEP patterns in BD (Fig. 5.5)[257]:

- No identifiable waves (73.34%)
- An isolated bilateral wave I (16.66%)
- An isolated unilateral wave I (10.00%)
- Waves II, III, IV, and V were not observed in any of the cases.

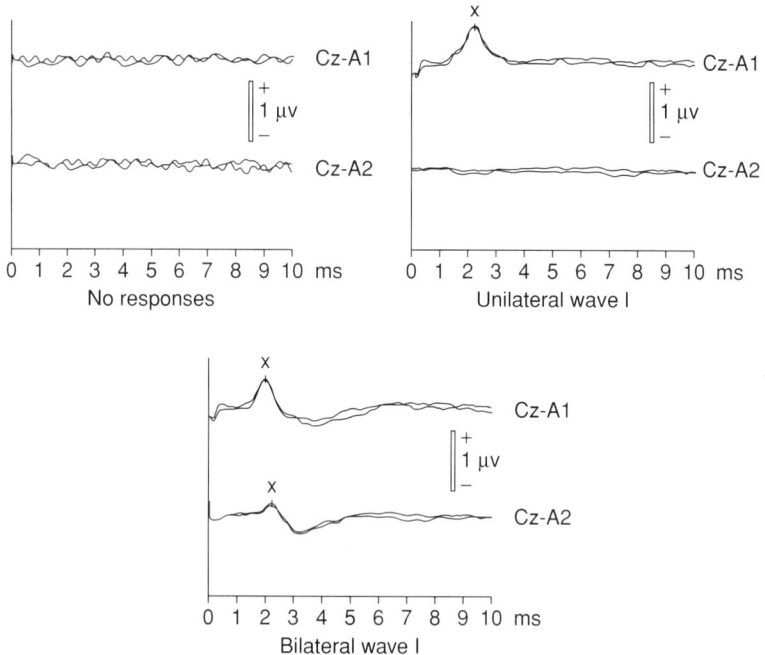

FIGURE 5.5. Brainstem auditory evoked potential (BAEP). (Modified from reference # 275.)

The detection of wave I, without any later components, is the most clear-cut BAEP finding compatible with BD.[257,335,337,338] However, a higher proportion of cases with bilateral absence of responses has been reported by most authors: 59% by Starr,[328] 77% by Goldie et al.,[330] 55% by Chen,[338] and 63.2% by Paniagua-Soto et al.[336] This BAEP pattern does not contribute to BD diagnosis because, when the history is incomplete, it is often not possible to exclude those patients who may have had preexisting deafness.[330] In head trauma, transverse fracture of the temporal bone could damage the cochlea or the statoacoustic nerve or produce hemotympanum.[291,330,339–345] The presence of only wave I, attributed to the auditory nerve action potential near the spiral ganglion,[328,346–351] is probably related to the interruption of intracranial circulation, which would interfere with the blood supply to the eighth nerve and cochlea, via the internal auditory artery.[257,330,352–359]

Whatever the reason for its presence in some BD patients, wave I in this clinical state is obviously different from that recorded in normal subjects. Its major difference consists of an increased latency. We found that wave I was significantly delayed in all cases with preservation of this BAEP component. Significant augmentation of wave I amplitude was present bilaterally in one case, and unilaterally in another. Wave I amplitude was significantly diminished in two patients.[40,257]

Additional evidence was provided by using serial BAEP records, which was possible in a few cases. It seems that the progressive hypoxic-ischemic damage requires several hours for complete dysfunction of the cochlea and the statoacoustic

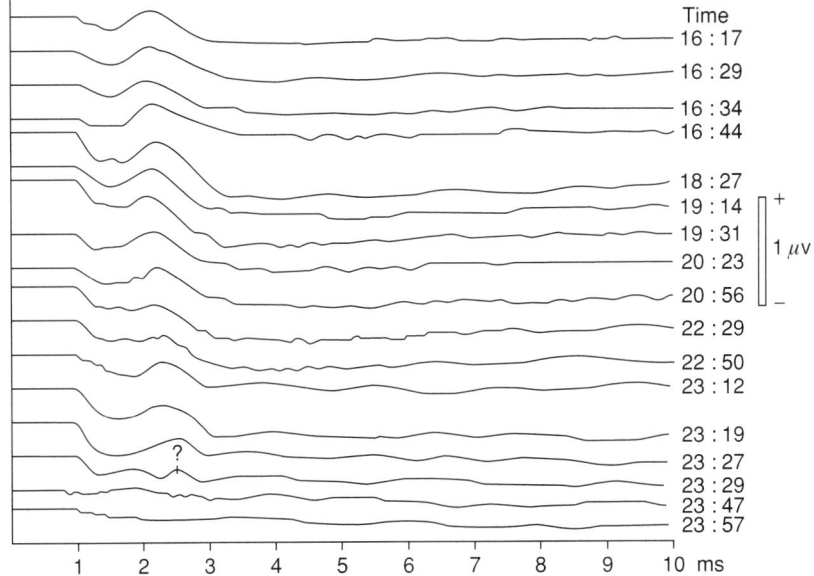

FIGURE 5.6. Wave I increases in latency progressively, over a period of up to 7 hours, until this BAEP disappears. (Modified from reference # 275.)

nerve, because in three of our cases wave I increased in latency progressively, over a period of up to 7 hours, after which this component disappeared. Other authors have also found a progressive disappearance of wave I, using serial BAEP recording (Fig. 5.6).[77,331,332,334,338,360,361]

Uziel et al.[362–364] reported a BD patient, with absence of all BAEP components, in whom transtympanic electrocochleography showed reduction in amplitude of N1 but revealed preservation of the cochlear microphonic potential. Other authors have reported similar observations.[365–368]

Furthermore, in one of our patients in whom serial BAEP records were done, a significant negative correlation was found for wave I latency with body temperature and heart rate. Several investigators have reported an increase of latency for all BAEP components in hypothermia.[331,332,369–375] It is well known that when corporeal temperature decreases, conduction velocity in sensory pathways is also reduced.[376] This may account for the significant correlation between wave I latency and body temperature in one of our patients. Thus, two facts might explain the progressive delay and disappearance of wave I in BD patients: progressive hypoxic-ischemic dysfunction of the cochlea and the eighth nerve plus hypothermia, often present in BD patients because of the failure of the thermoregulatory mechanisms. Hence, it can be suggested, that the incidence of wave I preservation reported by different authors based on single records might depend on the time elapsed between the moment when the condition of BD was initiated and the evoked potential study.[36,37,223,284]

It is more difficult to explain the finding of a significant correlation between wave I latency and heart rate in one of our cases, because it has been emphasized that blood supply to the cochlea and the statoacoustic nerve derives only from the vertebrobasilar system,[291,332,355,356,358,377-383] and in those patients who fulfilled the clinical and EEG criteria of BD, a cessation of all intracranial circulation should be expected.[140,152,154,279,384-386]

Nevertheless, heart rate and body temperature were also significantly correlated; that is, when body temperature decreased, heart rate also diminished.[257] On the other hand, several facts are considered to explain heart rate regulation in normal persons and in comatose and BD patients.[138,387-402] Further investigation is needed on this subject.

The increase of wave I amplitude in some of our BD patients warrants mention. Starr[328] first described substantial fluctuations of wave I amplitude, which has also been reported in severe brainstem dysfunction and barbiturate coma.[331,332,335] We also found fluctuations of wave I amplitude during serial BAEP recording after BD was established. Then, whether or not augmentation of wave I amplitude is found in single BAEP records might depend on the moment at which the evoked potential study is done in relation to the duration of the clinically BD state. Perhaps the unusually large amplitude of wave I in BD reflects a loss of centrifugal central nervous system (CNS) inhibition of cochlear activity.[335,375,403-405] However, changes of intracranial impedance, due to variations of CBF until all intracranial circulation is ended in BD, should be considered as a possible explanation for these interesting fluctuations of wave I amplitude.[257,406-409]

Wave II seems to be generated in the proximal part of the VIII nerve, probably at the cerebrospinal fluid (CSF)-brainstem interface, and in cochlear nuclei, with a mixed peripheral and central origin.[329,410,411] Some authors have reported the presence of wave II in their series of BD patients.[334,336,412] We recorded a unilateral wave II in a case with electrocerebral silence and no brainstem reflexes, but during the apnea test a spontaneous inspiration was observed, and thus the BD diagnosis was not confirmed. Two hours later, the apnea test was negative, but only a bilateral wave I was recorded at that moment.[257,353] The presence of wave II may give rise to some uncertainty in BD due to its dual origin, absence of the central component of which might perhaps be reflected in changes in latency and amplitude. However, the persistence of wave II seems to be of short duration in the progression toward BD, and at present it seems wiser to exclude its persistence from the certain patterns of BD.[334,336,412] The presence of wave II in BD patients may be evidence that at least part of this BAEP component arises from the proximal portion of the eighth nerve.[334,364,413]

As we have previously pointed out, the preservation of an isolated wave I has been taken as the most unequivocal BAEP pattern for the diagnosis of BD.[328,346-351] However, Ferbert et al.[295,414] reported the frequent occurrence of BAEPs with only wave I preserved, in patients with basilar artery thrombosis who still showed signs of brainstem function, suggesting that the BAEP technique is not useful for BD diagnosis in these cases and possibly in other brainstem lesions. Absent BAEPs have also been reported in patients with basilar artery thrombosis.[302,306] On the other hand, in isolated brainstem death, VEPs and the EEG can show preserved electrical function of the forebrain hemispheres, whereas BAEP and SEP findings resemble those found in BD.[229]

Nevertheless, BAEPs maintain all their value in the exclusion of false positives (that is, comatose patients looking brain dead), since a bilateral absence of response unrelated to BD is uncommon (deafness, ototoxic drug administration, and bilateral temporal bone fractures must be taken into account).[223] It may transiently occur in postanoxic coma (as may a flat EEG),[383,415] where some caution is to be used in their interpretation. Flat BAEPs may be observed in basilar artery occlusion as well.[306] In contrast, absent brainstem reflexes and flat EEG with preserved evoked potentials have been found by Facco et al.[416] and others,[417] illustrating their usefulness in identifying false-positive clinical diagnoses.[223] Similar diagnostic problems have been reported in encephalitis.[418–420]

As far as generators are concerned, it is well known that waves III to V are generated in the brainstem, and therefore their presence *does* exclude BD.[257,329,332,353,411,421]

Another important application of BAEPs is to monitor comatose patients. In Figure 5.7, BAEPs serial records are shown from a comatose patient who finally was brain dead.

Short Latency Somatosensory Evoked Potentials

A wealth of data regarding SEPs in BD are now available in the literature.[35,36,39,40,236,241,295,296,306,320–323,326,330,333,383,416,422–450] Unfortunately, several papers use a frontal reference, which prevents a proper recording of far-field potentials, an essential parameter in BD. This aspect calls for a short description of generators of cervical and far-field SEP components, which can be summarized as follows[447,448,451–457]:

- The brachial plexus for the cervical N9 and the far-field, scalp-recorded P9
- The ascending volley in the dorsal columns of the cervical spinal cord for the cervical N11 and the far-field, scalp-recorded P11
- The dorsal horn in the cervical spinal cord (with a dipole perpendicular to the spinal cord axis) for the cervical N13
- The higher cervical segment or cervicomedullary junction for far-field scalp-recorded P13
- The ascending volley in the caudal medial lemniscus for the P14.

Concerning the N18, studies performed in the past two decades have allowed for considerable clarification of its generator. Desmedt and Cheron[458] suggested that N18 probably reflected subcortical activities such as delayed potentials in the thalamus itself, or action potentials in the lower part of thalamocortical axons. For Mauguière et al.,[459] N18 is not a unitary phenomenon, but reflects several deeply located generators on the brainstem or thalamus. Urasaki et al.[460] suggested that N18 comes from brainstem activity between the upper pons and the midbrain rather than from the thalamus, and Sonoo et al.[455,456] concluded that

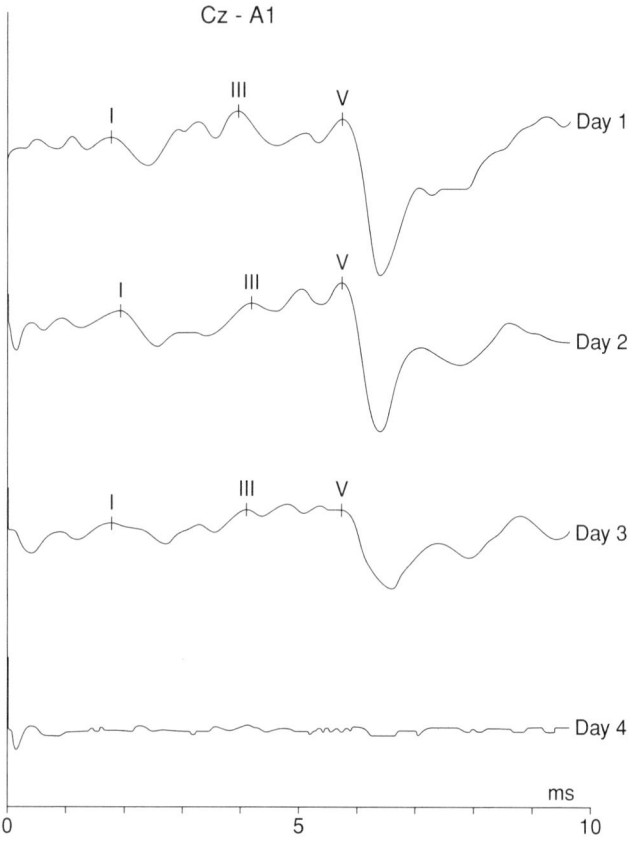

FIGURE 5.7. Serial BAEP records in a comatose patient. A progressive delay and deterioration of waves are shown until no responses are recorded on the fourth day.

at least a significant part of the N18 potential is generated caudal to the pontine level (probably in the medulla oblongata), or at higher levels via extralemniscal ways. Tomberg et al.[461] also documented a medullary origin for the N18 far-field potential.

A proper SEP recording in BD cannot but use a noncephalic reference, in order to detect the far-field components P13, P14, and N18.[38–40,223] The use of a frontal reference, which had been widely used in the past, cancels out all far-field components and allows for the assessment of cortical function only, making SEPs similar to EEG in this context. One can only tell whether the activity of somatosensory cortex is preserved or not, a small thing in the diagnosis of BD. On the other hand, SEPs with a noncephalic reference provide very useful information about the caudal part of the brainstem. Thus, the combined use of BAEPs and SEPs allows assessment of pontomesencephalic and bulbar levels, respectively, with a good capability of checking the level of rostrocaudal deterioration in preterminal states, down to brain death.[416] A nasopharyngeal derivation may be added as well, to get a near-field recording of the caudal and rostral components of P14.[218,451,462]

The SEPs demonstrate arrest of conduction at the cervicomedullary level in most patients, about 95% in our experience, while in the remaining cases, the cervical components are not recordable[223]; like flat BAEPs, their absence depends on extracerebral damage, preventing the evaluation of central pathways. The main causes of flat SEPs are lesions in the brachial plexus and cervical spinal cord as well as polyradiculopathies.[420] Much caution is to be used in high cervical vertebral fractures, where a present cervical N13 with absent P14, resembling a BD pattern, only shows the arrest of conduction in the spinal cord.[463] Since SEPs allow checking the caudal brainstem, their far-field potentials may still be preserved when brainstem reflexes and auditory brainstem response (ABR) waves II to V are already absent, and a close relationship between the onset of apnea and disappearance of N18 may be found.[416,456]

The SEP patterns of BD are as follows[39,40,60,223,288,416,454,456,464–470]:

- Absence of all components following N13/P13 components
- The dissociation between N13/P13 and P14, that is, the persistence of cervical N13/P13 with absent P14
- Regarding N18, considering the medullary origin of N18, this component helps to avoid premature apnea testing; in fact, the apnea test may be dangerous in comatose, non-BD patients, and SEP recording may allow postponing it until the disappearance of N18.

Hence, the preservation of P14 or N18 SEP components could indicate that the lower part of the brainstem is still functioning, and that a definitive BD diagnosis must be postponed until future examinations.[471]

It is also important to describe the SEP patterns in BD when using restricted filter bandpass setting, with special emphasis on the dissociation of some SEP components when cephalic and noncephalic references are used, because some investigators have emphasized that when filtering frequencies below 100 Hz, some subcortical components are better visualized and more precise clinical use of short-latency somatosensory responses is possible (Fig. 5.8).[472]

A characteristic SEP pattern was found: absence of Nm and later responses in the Cc′-Fpz lead and preservation of all or some of the so-called subcortical components in the rest of the derivations. We noted an interesting dissociation in which some SSEP components were not recorded in the scalp-noncephalic derivations, but they were still present in the neck-cephalic and spine channels, with inverse polarity. We selected a restricted bandpass for the scalp-noncephalic, neck-cephalic, and spine derivations, and we found this dissociation not only for P13-N13 and P14-N14 but for Nm-Pm′ and Pm-Nm′ components. It is likely that components Nm-Pm′ and Pm-Nm′ are "far-field potentials" that appear when the slow wave N18 is filtered out by the use of a restricted low filter setting (Fig. 5.9).

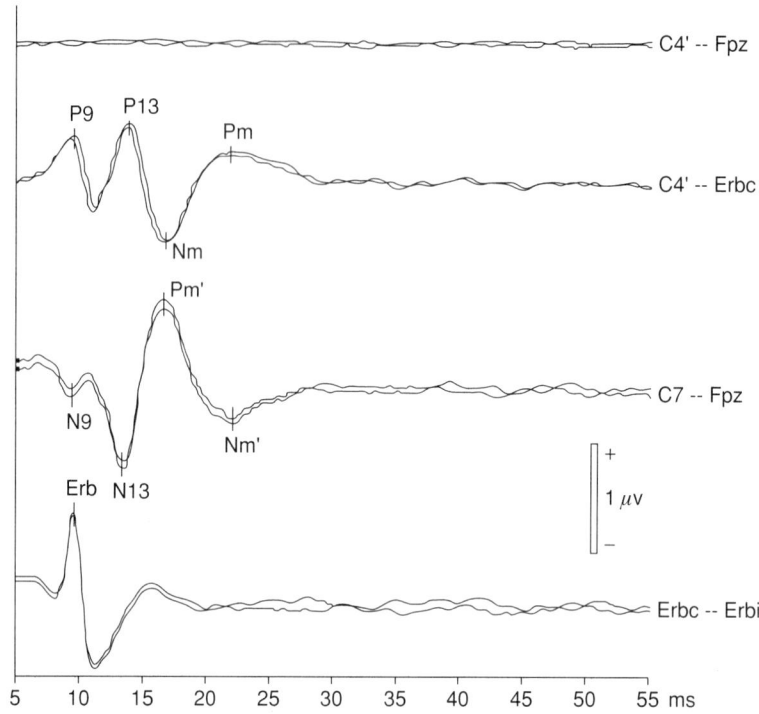

FIGURE 5.8. A characteristic somatosensory evoked potential (SEP) pattern was found: absence of Nm and later responses in the Cc'-Fpz lead and preservation of all or some of the so-called subcortical components of spinal and brachial plexus origin in the rest of the derivations. (Modified from reference # 40.)

Our data further suggest that all components alter N9-P9, recorded with restricted filter bandpass, and recognize two distinct generator sources. Rostral generators are located probably in the brainstem or thalamus, but at least a significant part of these SEP components are generated at the very lower part of the medulla oblongata (dorsal column nuclei) or at the upper cervical spine. Serial SEP records let us affirm that the presence or absence of this dissociation in single SEP records of BD patients could depend on the moment when the evoked potential study is done in relation to BD evolution.[37–40,223,354,418,473,474]

The SEPs are also very useful to monitor comatose patients during their clinical evolutions (Fig. 5.10).

Multimodality Evoked Potentials

Apart from the specific advantages and limitations of each modality, it is obvious that MEPs provide a better assessment of BD compared to any single modality,

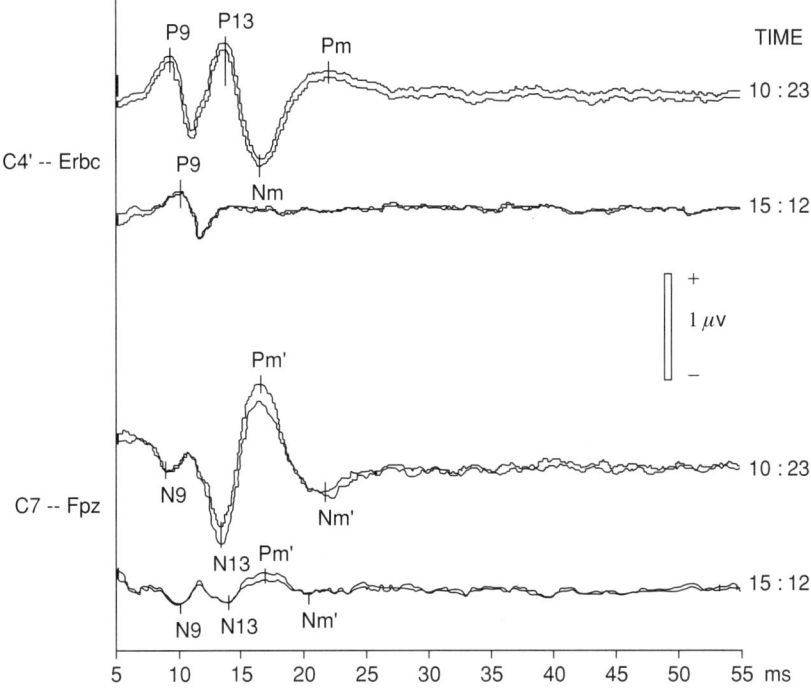

FIGURE 5.9. We selected a restricted bandpass for the scalp-noncephalic, neck-cephalic, and spine derivations, and we found this dissociation not only for P13-N13 and P14-N14 but for Nm-Pm' and Pm-Nm' components. (Modified from reference # 40.)

allowing assessment of different nervous pathways with different anatomic locations of their generators. The VEPs explore fronto-occipital hemispheric structures, the BAEPs the pons and mesencephalon, and the SEPs a long rostrocaudal path from parietal cortex to the cervical spinal cord (with relevant caudal generators in the medulla oblongata). Therefore, depending on the site and extent of primary and secondary lesions, a single modality might exclude BD. For example, VEPs might be preserved in brainstem lesions that extinguish both BAEPs and SEPs; conversely, patients with medullary lesions may retain BAEPs, while in patients with hemispheric lesions and rostrocaudal evolution, SEPs alone may disclose a still viable brainstem with reserved P14 or N18.[35,37,40,60,223,295,353,354,416,418,475–477]

When VEPS and ERG are elicited and recorded simultaneously, using cephalic and noncephalic references in BD, a clear result is found, confirming that in the visual pathways of BD patients, electrical activity is confined to the retina.[35,36,40,298,299] Moreover, the detection of wave I, without any later components, is the most clear-cut BAEP finding compatible with BD.[257,335,337,338] However, a higher proportion of cases with bilateral absence of responses has been reported by most authors. This BAEP pattern, as a single test, can provide

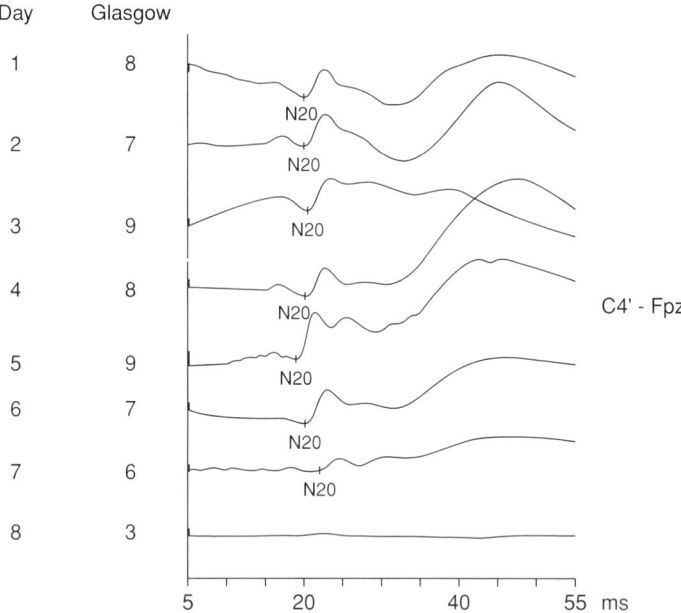

FIGURE 5.10. Serial SEP records in a comatose patient. Changes in wave 20 latency and amplitude correlate with the Glasgow Coma Scale. A progressive delay and deterioration of waves are shown until no responses are recorded on the eighth day.

a false-positive BD diagnosis because when the history is incomplete it is not possible to exclude that the patients may have had preexisting deafness.[257,330] As has been previously discussed, in head trauma, transverse fracture of the temporal bone could damage the cochlea or statoacoustic nerve or produce hemotympanum.[291,330,339–345]

Furthermore, MEPs enable the exclusion of BD in sedated patients, easily and noninvasively, therefore helping to optimize the timing of contrast angiography; in a few patients with a clinical and EEG picture of BD and no reversible factors, MEPs may even show residual function in the brainstem, illustrating how MEPs may also improve diagnostic safety.[416] These data once again emphasize that BD requires a careful diagnosis, regarding evaluation of the patient's history, clinical data, and appropriate use and interpretation of confirmatory tests, with no a priori choice of a single investigation. It seems odd nowadays that some published criteria still recommend EEG while underestimating, or not even mentioning, MEPs.[37,416]

Hence, we propose the use of a test battery, including EEG, BAEPs, short-latency SEPs, VEPs, and ERG to study BD patients. Such a test battery would permit the assessment of several sensory pathways and the evaluation of both brainstem and cerebral hemispheric functions. Thus, the reliability of the BD diagnosis could be considerably increased.[35,37–40,223,257,298–300,353,354,360,418,474,476,478]

Heart Rate Variability

Another interesting test that is not usually included as an ancillary test in BD confirmation is the measurement of heart rate variability.[388]

Heart Rate Variability Calculation

In a continuous ECG record, each QRS complex is detected, and the so-called normal-to-normal (N-N) intervals (that is, all intervals between adjacent QRS complexes resulting from sinus node depolarizations) or the instantaneous heart rate is determined. Other authors also characterize the N-N intervals as R-R intervals, because usually the R wave from the electrocardiogram is detected to measure these intervals.[479,480]

Instantaneous heart rate is not steady but rather demonstrates continuous fluctuations. As these oscillations in instantaneous heart rate depend largely on interactions between sympathetic and parasympathetic efferent activities, heart rate variability (HRV) analysis has been widely used as a measure of activity in the two components of the autonomic system. The measurement of HRV as the variation of the R-R intervals in the electrocardiogram has been widely used as a test of autonomic cardiovascular dysfunction. Heart rate after parasympathetic stimulation is dependent on the moment in the heart cycle when the stimulus occurs. As a result, heart rate is an irregular, discontinuous function of the vagal efferent impulse frequency.[389,394-397,481]

In the analysis of HRV, there are two methods: the time domain analysis and the frequency domain analysis. Time domain analysis allows measuring global sympathetic and parasympathetic HRV as does the frequency analysis. Frequency analysis allows separating the two major components (sympathetic and parasympathetic).[388-390,394-397,400,402]

Simple time domain variables are the mean R-R interval, the mean heart rate, the difference between the longest and shortest R-R interval, and the difference between night and day heart rate. Other common time domain measurements are in instantaneous heart rate secondary to respiration, tilt, Valsalva maneuver, or phenylephrine infusion. These differences can be described as differences in either heart rate or cycle length.[479,480]

The simplest variable to calculate HRV is the standard deviation of the R-R (SDNN) intervals, that is, the square root of variance. Since variance is mathematically equal to the total power of spectral analysis, SDNN reflects all the cyclic components responsible for variability. It is usually calculated over a period of 24 hours. Other commonly used statistical variables calculated from segments of the total monitoring period include SDARR, the standard deviation of the average R-R intervals calculated over short periods, usually 5 minutes, which is an estimate of the changes in heart rate due to cycles longer than 5 minutes, and the SDNN index, the mean of the 5-minute standard deviations of N-N intervals calculated over 24 hours, which measures the variability due to cycles shorter than 5 minutes.[389,479-480,481,482]

Other geometric methods have been developed in which the series of R-R intervals also can be converted into a geometric pattern such as the sample density distribution of N-N interval, durations, sample density distribution of differences between adjacent N-N intervals, and Lorenz plot of N-N or R-R intervals. According to the Task Force,[479,480] the time domain HRV analysis can be summarized as follows:

- SDRR (estimate of overall HRV)
- HRV triangular index (estimate of overall HRV)
- SDARR (estimate of long-term components of HRV)
- RMSSD (estimate of short-term components of HRV).

Power spectral analysis (PSA) of HRV has been applied for more than two decades.[483–489] It has been used for the evaluation of autonomic nervous functions following sympathetic or parasympathetic blockade.[490,491] Three main spectral components are distinguished in a spectrum calculated from short-term recordings of 2 to 5 minutes.[492–497] Long-term calculations of 24 hours or more also have been used.[541]

The standard procedure, definition, and interpretation had not been well established until 1996 (Task Force of the European Society of Cardiology and the North American Society of Pacing and Electrophysiology). The measurement of very low frequency (VLF), low frequency (LF), and high frequency (HF) power components is usually made in absolute values of power (milliseconds squared). The LF and HF may also be measured in normalized units, which represent the relative value of each power component in proportion to the total power minus the VLF component. The representation of LF and HF in normalized units emphasizes the controlled and balanced behavior of the two branches of the autonomic nervous system. Moreover, the normalization tends to minimize the effect of the changes in total power on the values of LF and HF components (Fig. 5.11).[479,480]

The distribution of the power and the central frequency of LF and HF are not fixed but may vary in relation to changes in autonomic modulations of heart period.[492–497] The physiological explanation of the VLF component is much less defined, and the existence of a specific physiological process attributable to these heart period changes might even be questioned. The nonharmonic component, which does not have coherent properties and is affected by algorithms of baseline or trend removal, is commonly accepted as a major constituent of VLF. Thus, VLF assessed from short-term recordings (5 minutes) is a dubious measure and should be avoided when the HRV of short-term ECGs is interpreted.[479,480]

The LF, which indicates the joint contribution of sympathetic and parasympathetic drives, was defined at a frequency range of 0.04 to 0.15 Hz. The HF, which indicates activities of the respiratory sinus arrhythmia (RSA) and vagus nerves, was identified at 0.15–0.4 Hz[479,480,492–497]:

- VLF: 0.0033–0.04 Hz
- LF: 0.04–0.15 Hz
- HF: 0.15–0.4 Hz.

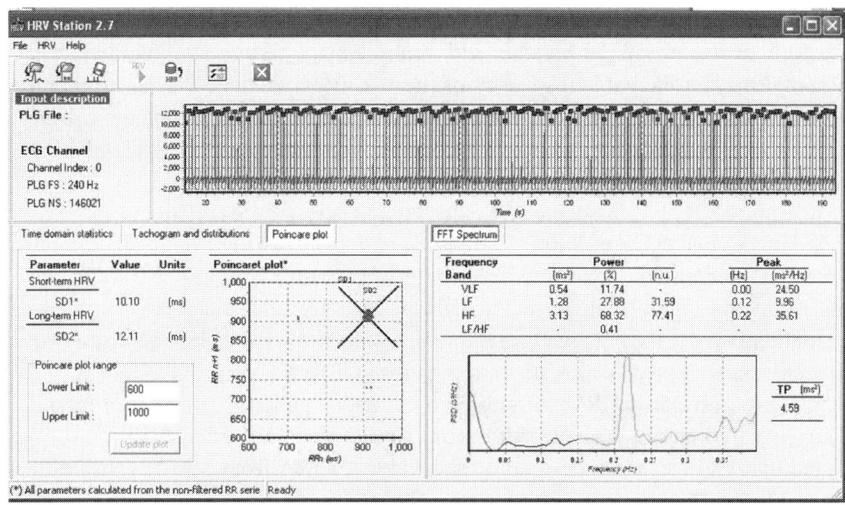

FIGURE 5.11. Heart rate variability in a brain-dead patient. Recorded by the heart rate variability (HRV) station system developed at the "Universidad Central de las Villas" laboratory. (See also color insert.)

Heart Rate Variability in Brain Death and Coma

Several papers have appeared in recent decades on the assessment of HRV in BD and coma. Schwarz et al.,[499] using a time domain analysis, found that a bivariate distribution of heart rate variability and of heart rate showed a discrimination of the group of comatose patients from BD subjects and normals. These authors concluded that an HRV <2.3% in BD is interpreted as a cerebrocardial dissociation. Freitas et al.[401,500] performed a linear discriminant analysis of HRV parameters between BD and deep coma patients, and reported that all patients in the data set were correctly classified (sensitivity and specificity of 100%).

Several authors have emphasized that HRV could be helpful as a predictor of imminent BD and a useful adjunct for predicting the outcome of patients with severe head injuries.[394,400,501–506] Rapenne et al.[398,402] observed comatose patients for 6 hours before BD, a progressive extinction of the influence of the autonomic nervous system on cardiovascular regulation, and found that there were no parasympathetic or sympathetic activities as soon as BD occurred.

Biswas et al.[393,400] found that children with a Glasgow Coma Scale (GCS) score of 3 to 4 had a lower LF/HF ratio compared with those who had a GCS score of 5 to 8, and that patients who progressed to BD had a markedly lower LF/HF ratio, with a significant decrease after the first 4 hours of hospitalization. Other patients with more favorable outcomes had significantly higher LF/HF ratios. These authors concluded that the LF/HF ratio may be helpful not only in identifying those patients who will progress to BD but also in predicting which patients will have favorable outcome.

Su et al.[389] correlated the parameters derived from spectral analysis of HRV with the GCS in five groups of patients with brain damage of various severities. They reported that an increase in severity of the brainstem damage was accompanied with an augmentation of a sympathetic and a decrement of parasympathetic drives, as indicated by increasing LF% and LF/HF and decreasing HF, respectively. Both LF and HF components were nearly abolished in BD.

Kawamoto et al.[387] assessed HRV during apnea tests in BD patients and found that $PaCO_2$ and LF ($p < .05$) increased, and pH ($p < .01$) and HF/LF ratio ($p < .05$) decreased. Heart rate, mean arterial pressure, PaO_2, and HF remained consistent throughout. They concluded that sympathovagal balance was inclined to be sympathotonic during apnea, and that there were no changes in the respiratory-related vagal activity in spite of stopping artificial ventilation.

We also correlated HRV with GCS in comatose patients using serial records. Thirty-five comatose patients with acute stroke were studied serially.[388,481] We found that GCS scores increased or decreased ranging from 3 to 15, with either improvement or deterioration of coma. Patients were divided into four groups (GCS from 10 to 15; GCS from 7 to 9; GCS from 3 to 7; and BD). An HRV was calculated in the time domain according to the following formula:

$$HRV = (SD/M) \times 100$$

where SD is standard deviation
M is heart rate mean

Forty normal subjects were also studied. We found that HRV did not progressively decrease when GCS scores diminished from 15 to 3, according to patients' evolution. The HRV mean values and dispersion were highest in the group with a GCS score ranging from 7 to 9. A best-fitting curve for a nonlinear correlation model of HRV and GCS also showed a statistically significant increase of HRV in this group of patients. A GCS score of about 8 has been related to a diencephalic level of consciousness impairment.[478] Hence, this remarkable, previously unreported observation of HRV increasing when supratentorial brain damage causes dysfunction at the diencephalic level might be related to functional suppression of the hypothalamic defense area with retention of brainstem integrity in coma (Fig. 5.12).

Recent reports have described a complex and distinctly nonlinear heart rhythm in adults and in mature infants.[507,508] However, Sugihara et al.[508] have found that such nonlinearity is lacking in preterm infants (gestation ≥ 27 weeks) where parasympathetic-sympathetic interaction and function are presumed to be less well developed. These authors emphasized that BD infants and those treated with atropine exhibit a similar lack of nonlinear feedback control.

In conclusion, HRV is a powerful tool for assessing autonomic function in comatose patients, with a high predictive value of clinical evolution. It also could be considered as an additional investigation in a battery of ancillary tests to confirm the diagnosis of BD.[284,388,481]

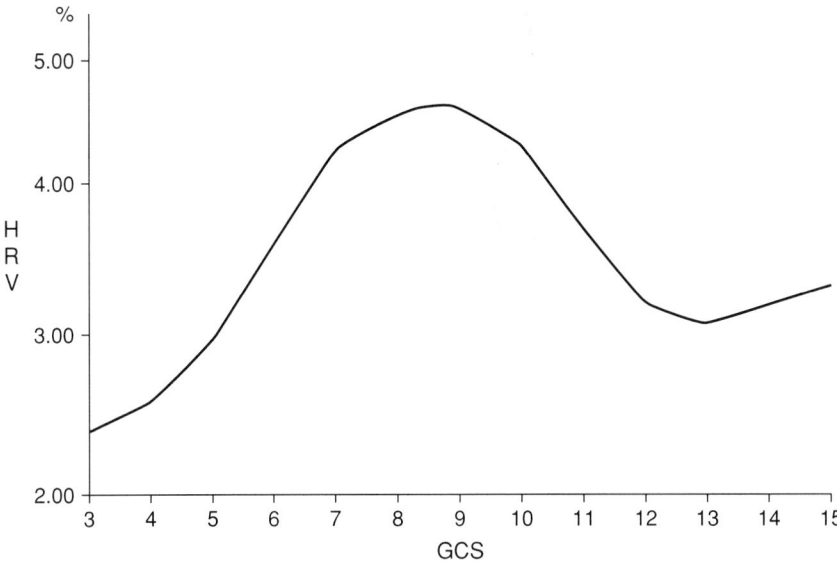

FIGURE 5.12. We found that HRV did not progressively decrease when Glasgow Coma Scale (GCS) scores diminished from 15 to 3, according to the patients' evolution. HRV mean values and dispersion were highest in the group with a GCS score ranging from 7 to 9. A best-fitting curve for a nonlinear correlation model of HRV and GCS also showed a statistically significant increase of HRV in this group of patients. (Modified from reference # 524.)

References

1. Wijdicks EF. Determining brain death in adults. *Neurology.* 1995;45:1003–1011.
2. Wijdicks EF. The first organ transplant from a brain-dead donor. *Neurology.* 2006;66:460–461.
3. Wijdicks EF, Cranford RE. Clinical diagnosis of prolonged states of impaired consciousness in adults. *Mayo Clin Proc.* 2005;80:1037–1046.
4. Wijdicks EF. The diagnosis of brain death. *N Engl J Med.* 2001;344:1215–1221.
5. Wijdicks EFM. Clinical Diagnosis and Confirmatory Testing in Brain Death in Adults. In: Wijdicks EFM, ed. *Brain Death.* Philadelphia: Lippincott Williams & Wilkins, 2001:61–90.
6. Ashwal S. Clinical diagnosis and confirmatory testing of brain death in children. In: Wijdicks EFM, ed. *Brain Death.* Philadelphia: Lippincott Williams & Wilkins, 2001:91–114.
7. Walker AE. *Cerebral Death,* 3rd ed. Baltimore-Munich: Urban & Schwarzenberg, 1985.
8. Walker AE. Current concepts of brain death. *J Neurosurg Nurs.* 1983;15:261–264.
9. An appraisal of the criteria of cerebral death. A summary statement. A collaborative study. *JAMA.* 1977;237:982–986.

10. Duran-Ferreras E, Duran-Ferreras A, Redondo-Verge L, Castro-Montano J, Varez-Marquez E, Rodriguez-de Quesada B. [When should a brain scan with HMPAO be performed to diagnose brain death?]. *Rev Neurol.* 2003;36:941–943.

11. Saito T, Kurashima A, Oda T, et al. [Quantitative analysis of plasma concentration of barbiturate for diagnosis of brain death]. *No Shinkei Geka.* 2002;30:593–599.

12. Lopez-Navidad A, Caballero F, Domingo P, et al. Early diagnosis of brain death in patients treated with central nervous system depressant drugs. *Transplantation.* 2000;70:131–135.

13. Schwab S, Spranger M, Schwarz S, Hacke W. Barbiturate coma in severe hemispheric stroke: useful or obsolete? *Neurology.* 1997;48:1608–1613.

14. Saito T, Takeichi S, Tokunaga I, Nakajima Y, Osawa M, Yukawa N. Experimental studies on effects of barbiturate on electroencephalogram and auditory brain-stem responses. *Nippon Hoigaku Zasshi.* 1997;51:388–395.

15. Link J, Schaefer M, Lang M. Concepts and diagnosis of brain death. *Forensic Sci Int.* 1994;69:195–203.

16. Rabanal JM, Teja JL, Quesada A, Fleitas MG, Reganon GD, Herrera S. Do barbiturates in brain-dead organ donors have a deleterious effect on early renal graft function? *Transplant Proc.* 1991;23:2492.

17. LaMancusa J, Cooper R, Vieth R, Wright F. The effects of the falling therapeutic and subtherapeutic barbiturate blood levels on electrocerebral silence in clinically brain-dead children. *Clin Electroencephalogr.* 1991;22:112–117.

18. Toffol GJ, Lansky LL, Hughes JR, et al. Pitfalls in diagnosing brain death in infancy. *J Child Neurol.* 1987;2:134–138.

19. Powner DJ. Drug-associated isoelectric EEGs. A hazard in brain-death certification. *JAMA.* 1976;236:1123.

20. Gurvitch AM. Determination of the depth and reversibility of post-anoxic coma in animals. *Resuscitation.* 1974;3:1–26.

21. Fermaglich JL. Determining cerebral death. *Am Fam Physician.* 1971;3:85–87.

22. Wijdicks EF. Brain death worldwide: accepted fact but no global consensus in diagnostic criteria. *Neurology.* 2002;58:20–25.

23. Qureshi AI, Kirmani JF, Xavier AR, Siddiqui AM. Computed tomographic angiography for diagnosis of brain death. *Neurology.* 2004;62:652–653.

24. Tsai WH, Lee WT, Hung KL. Determination of brain death in children—a medical center experience. *Acta Paediatr Taiwan.* 2005;46:132–137.

25. Okuyaz C, Gucuyener K, Karabacak NI, Aydin K, Serdaroglu A, Cingi E. Tc-99m-HMPAO SPECT in the diagnosis of brain death in children. *Pediatr Int.* 2004;46:711–714.

26. Mejia RE, Pollack MM. Variability in brain death determination practices in children. *JAMA.* 1995;274:550–553.

27. Paret G, Barzilay Z. Apnea testing in suspected brain dead children—physiological and mathematical modelling. *Intensive Care Med.* 1995;21:247–252.

28. Chantarojanasiri T, Preutthipan A. Apnea documentation for determination of brain death in Thai children. *J Med Assoc Thai.* 1993;76(suppl 2):165–168.

29. Galaske RG, Schober O, Heyer R. Determination of brain death in children with 123I-IMP and Tc-99m HMPAO. *Psychiatry Res.* 1989;29:343–345.

30. Machado C. Determination of death. *Acta Anaesthesiol Scand.* 2005;49:592–593.

31. Machado C, Abeledo M, Alvarez C, et al. Cuba has passed a law for the determination and certification of death. *Adv Exp Med Biol.* 2004;550:139–142.

32. Machado C, Shewmon DL. *Brain Death and Disorders of Consciousness.* New York: Kluwer Academics/Plenum, 2004.

33. Machado C, García OD. Guidelines for the determination of brain death. In: Machado C, ed. *Brain Death (Proceedings of the Second International Symposium on Brain Death).* Amsterdam: Elsevier Science BV, 1995:75–80.

34. Wijdicks EFM. *Brain Death.* Philadelphia: Lippincott Williams & Wilkins, 2001.

35. Machado C. Evoked potentials in brain death. *Clin Neurophysiol.* 2004;115:238–239.

36. Machado-Curbelo C, Roman-Murga JM. [Usefulness of multimodal evoked potentials and the electroretinogram in the early diagnosis of brain death]. *Rev Neurol.* 1998;27:809–817.

37. Machado C. A contribution of multimodality evoked potentials and electroretinography for the early diagnosis of brain death. In: Machado C, ed. *Brain Death (Proceedings of the Second International Symposium on Brain Death).* Amsterdam: Elsevier Science BV, 1995:141–50.

38. Machado C. An early approach to brain death diagnosis using multimodality evoked potentials and electroretinography. *Minerva Anestesiol.* 1994;60:573–577.

39. Machado C, Valdes P, Garcia O, Coutin P, Miranda J, Roman J. Short latency somatosensory evoked potentials in brain-dead patients using restricted low cut filter setting. *J Neurosurg Sci.* 1993;37:133–140.

40. Machado C. Multimodality evoked potentials and electroretinography in a test battery for an early diagnosis of brain death. *J Neurosurg Sci.* 1993;37:125–131.

41. Settergren G. Brain death: an important paradigm shift in the 20th century. *Acta Anaesthesiol Scand.* 2003;47:1053–1058.

42. Löfstedt S, von Reis G. Intrakraniella lesioner med bilateralt upphävd kontrastpassage i a. carotis interna (Intracranial lesions with abolished passage of x-ray contrast through the internal carotid arteries). *Opusc Med.* 1956;1:199–202.

43. Löfstedt S von. Diminution or obstruction of blood flow in the internal carotid artery. *Opusc Med.* 1959;4:345–360.

44. Wertheimer P, de Roug, Descotes J, Jouvet M. [Angiographical data concerning the death of the brain during comas with respiratory arrest (so-called protracted coma)]. *Lyon Chir.* 1960;56:641–648.

45. Wertheimer P, Jouvet M, Descotes J. A propos du diagnostic de la mort du système nerveux dans les comas avec arrêt respiratoire traites par respiration artficielle. *Presse Med.* 1959;67:87–88.

46. Moussouttas M, Giacino J, Papamitsakis N. Amnestic syndrome of the subcallosal artery: a novel infarct syndrome. *Cerebrovasc Dis.* 2005;19:410–414.

47. Walker AE. Ancillary studies in the diagnosis of brain death. *Ann N Y Acad Sci.* 1978;315:228–240.

48. Kricheff II, Pinto RS, George AE, Braunstein P, Korein J. Angiographic findings in brain death. *Ann N Y Acad Sci.* 1978;315:168–183.

49. Black PM. Brain death (first of two parts). *N Engl J Med.* 1978;299:338–344.

50. Korein J, Braunstein P, George A, et al. Brain death: I. Angiographic correlation with the radioisotopic bolus technique for evaluation of critical deficit of cerebral blood flow. *Ann Neurol.* 1977;2:195–205.

51. Robert F, Mumenthaler M. [Criteria of brain death. Spinal reflexes in 45 personal studies]. *Schweiz Med Wochenschr.* 1977;107:335–341.

52. Korein J. Neurology and cerebral death—definitions and differential diagnosis. *Trans Am Neurol Assoc.* 1975;100:210–212.

53. Smith AJ, Walker AE. Cerebral blood flow and brain metabolism as indicators of cerebral death. A review. *Johns Hopkins Med J.* 1973;133:107–119.

54. Jørgensen EO. Aortocervical, internal carotid and vertebral angiography in total cerebral infarction. *Acta Radiol Diagn (Stockh).* 1973;14:369–378.

55. Jørgensen PB, Jørgensen EO, Rosenklint A. Brain death pathogenesis and diagnosis. *Acta Neurol Scand.* 1973;49:355–367.

56. Ouaknine G, Kosary IZ, Braham J, Czerniak P, Nathan H. Laboratory criteria of brain death. *J Neurosurg.* 1973;39:429–433.

57. Braunstein P, Korein J, Kricheff I. Bedside assessment of cerebral circulation. *Lancet.* 1972;1:1291–1292.

58. Ingvar DH, Widen L. [Brain death. Summary of a symposium]. *Lakartidningen.* 1972;69:3804–3814.

59. Balslev-Jørgensen P, Heilbrun MP, Boysen G, Rosenklint A, Jørgensen EO. Cerebral perfusion pressure correlated with regional cerebral blood flow, EEG and aortocervical arteriography in patients with severe brain disorders progressing to brain death. *Eur Neurol.* 1972;8:207–212.

60. Munari M, Zucchetta P, Carollo C, et al. Confirmatory tests in the diagnosis of brain death: comparison between SPECT and contrast angiography. *Crit Care Med.* 2005;33:2068–2073.

61. Young GB, Lee D. A critique of ancillary tests for brain death. *Neurocrit Care.* 2004;1:499–508.

62. Stulin ID, Sinkin MV. [Current clinical and instrumental diagnosis of brain death]. *Zh Nevrol Psikhiatr Im S S Korsakova.* 2006;106:58–64.

63. Brommeland T, Hennig R, Bajic R. [Should cerebral angiography still be required in organ donation?]. *Tidsskr Nor Laegeforen.* 2004;124:2513.

64. Haupt WF, Rudolf J. European brain death codes: a comparison of national guidelines. *J Neurol.* 1999;246:432–437.

65. Baron M, Brasfield J. A 37-year-old man with severe head trauma, and a "hot nose" sign on brain flow study. *Chest.* 1999;116:1468–1470.

66. Bernat JL. On irreversibility as a prerequisite for brain death determination. *Adv Exp Med Biol.* 2004;550:161–167.

67. Walker AE, Diamond EL, Moseley J. The neuropathological findings in irreversible coma. A critque of the "respirator". *J Neuropathol Exp Neurol.* 1975;34:295–323.

68. Perez-Pinzon MA, Dave KR, Raval AP. Role of reactive oxygen species and protein kinase C in ischemic tolerance in the brain. *Antioxid Redox Signal.* 2005;7:1150–1157.

69. Wang Q, Sun AY, Simonyi A, et al. Neuroprotective mechanisms of curcumin against cerebral ischemia-induced neuronal apoptosis and behavioral deficits. *J Neurosci Res.* 2005;82:138–148.

70. Ingvar DH. Brain death—total brain infarction. *Acta Anaesthesiol Scand Suppl.* 1971;45:129–140.

71. Juengling FD, Kassubek J, Huppertz HJ, Krause T, Els T. Separating functional and structural damage in persistent vegetative state using combined voxel-based analysis of 3-D MRI and FDG-PET. *J Neurol Sci.* 2005;228:179–184.

72. Yeh TH, Wang HL. Global ischemia downregulates the function of metabotropic glutamate receptor subtype 5 in hippocampal CA1 pyramidal neurons. *Mol Cell Neurosci.* 2005;29:484–492.

73. Coimbra CG. Implications of ischemic penumbra for the diagnosis of brain death. *Braz J Med Biol Res.* 1999;32:1479–1487.

74. Pallis C. Brain stem death—the evolution of a concept. *Med Leg J.* 1987;55(pt 2):84–107.

75. Pallis C. Whole-brain death reconsidered—physiological facts and philosophy. *J Med Ethics.* 1983;9:32–37.

76. Ingvar DH. Clinical neurophysiology of the cerebral circulation. *Electroencephalogr Clin Neurophysiol Suppl.* 1978;34:71–81.

77. Shiogai T, Saito I. [Ancillary studies in the determination of brain death]. *Nippon Rinsho.* 1997;55(suppl 1):301–309.

78. Ishii K, Onuma T, Kinoshita T, Shiina G, Kameyama M, Shimosegawa Y. Brain death: MR and MR angiography. *AJNR Am J Neuroradiol.* 1996;17:731–735.

79. Monsein LH. The imaging of brain death. *Anaesth Intensive Care.* 1995;23:44–50.

80. Picard L, Braun M, Anxionnat R, et al. [Venous angiography: importance in the diagnosis of brain death. 125 cases]. *Bull Acad Natl Med.* 1995;179:27–37.

81. Cantu RC. Brain death as determined by cerebral arteriography. *Lancet.* 1973;1:1391–1392.

82. Bergquist E, Bergstrom K. Angiography in cerebral death. *Acta Radiol Diagn (Stockh).* 1972;12:283–288.

83. Bergquist E, Bergstrom K. Angiographic documentation of the development of cerebral circulatory arrest. *Acta Radiol Diagn (Stockh).* 1972;12:7–11.

84. Korein J. The diagnosis of brain death. *Semin Neurol.* 1984;4:52–72.

85. Pallis C. Further thoughts on brainstem death. *Anaesth Intensive Care.* 1995;23:20–23.

86. Pallis C. ABC of brain stem death. Diagnosis of brain stem death–II. *Br Med J (Clin Res Ed).* 1982;285:1641–1644.

87. Pallis C. ABC of brain stem death. Diagnosis of brain stem death–I. *Br Med J (Clin Res Ed).* 1982;285:1558–1560.

88. Hazratji SM, Singh BM, Strobos RJ. Angiography in brain death. *N Y State J Med.* 1981;81:82–83.

89. Walker AE, Molinari GF. Criteria of cerebral death. *Trans Am Neurol Assoc.* 1975;100:29–35.

90. Louvier N, Combes JC, Nicolas F, Freysz M, Wilkening M. Cerebral angiography must have medicolegal value for brain death confirmation in France. *Transplant Proc.* 1996;28:377.

91. Gros C, Vlahovitch B, Frerebeau P, et al. [Arteriographic criteria of irreversible coma in neurosurgery]. *Neurochirurgie.* 1969;15:477–486.

92. Gros C, Vlatovitch B, Roilgen A. Les arrêts circulatoires dâns l'hypertension intracrânienne suraiguë. *Presse Med.* 1959;67:1065–1067.

93. Gros C, Vlatovitch B, Roilgen A. Images angiographiques d'arrêt circulatoire total dans les souffrances aiguës du tronc cérébral. *Neurochirurgie.* 1959;5:113–129.

94. Heiskanen O. Cerebral circulatory arrest caused by acute increase of intracranial pressure. A clinical and roentgenologica. Study of 25 cases. *Acta Neurol Scand.* 1964;40(suppl):7–57.

95. Troupp H, Heiskanen O. Cerebral angiography in cases of extremely high intracranial pressure. *Acta Neurol Scand.* 1963;39:213–223.

96. Bucheler E, Kaufer C. [Carotid and vertebral angiography in brain death]. *Acta Radiol Diagn (Stockh).* 1972;13:301–311.

97. Bucheler E, Kaufer C, Dux A. [Cerebral angiography to determine brain death]. *Fortschr Geb Rontgenstr Nuklearmed.* 1970;113:278–296.

98. Vlahovitch B, Frerebeau P, Kuhner A, Billet M, Gros C. [Angiographies under pressure in brain death with encephalic circulatory arrest]. *Neurochirurgie.* 1971;17:81–96.
99. Bradac GB, Simon RS. Angiography in brain death. *Neuroradiology.* 1974;7:25–28.
100. Rosenklint A, Jørgensen PB. Evaluation of angiographic methods in the diagnosis of brain death. Correlation with local and systemic arterial pressure and intracranial pressure. *Neuroradiology.* 1974;7:215–219.
101. [No authors listed]. How to determine brain death in adults: new guidelines. *J Crit Illn.* 1995;10(10):669–670.
102. Ringelstein EB, Zeumer H, Poeck K. Non-invasive diagnosis of intracranial lesions in the vertebrobasilar system. A comparison of Doppler sonographic and angiographic findings. *Stroke.* 1985;16:848–855.
103. Goodman JM, Heck LL, Moore BD. Confirmation of brain death with portable isotope angiography: a review of 204 consecutive cases. *Neurosurgery.* 1985;16:492–497.
104. Albertini A, Schonfeld S, Hiatt M, Hegyi T. Digital subtraction angiography—a new approach to brain death determination in the newborn. *Pediatr Radiol.* 1993;23:195–197.
105. Meguro K, Tsukada A, Matsumura A, Matsuki T, Nakada Y, Nose T. Portable digital subtraction angiography in the operating room and intensive care unit. *Neurol Med Chir (Tokyo).* 1991;31:768–772.
106. Ducrocq X, Braun M, Debouverie M, Junges C, Hummer M, Vespignani H. Brain death and transcranial Doppler: experience in 130 cases of brain dead patients. *J Neurol Sci.* 1998;160:41–46.
107. Nau R, Prange HW, Klingelhofer J, et al. Results of four technical investigations in fifty clinically brain dead patients. *Intensive Care Med.* 1992;18:82–88.
108. Van Bunnen Y, Delcour C, Wery D, Richoz B, Struyven J. Intravenous digital subtraction angiography. A criteria of brain death. *Ann Radiol (Paris).* 1989;32(4):279–281.
109. Braum M, Ducrocq X, Huot JC, Audibert G, Anxionnat R, Picard L. Intravenous angiography in brain death: report of 140 patients. *Neuroradiology.* 1997;39:400–405.
110. de Campo MP. Imaging of brain death in neonates and young infants. *J Paediatr Child Health.* 1993;29:255–258.
111. Mueller DL, Amundson GM, Wesenberg RL, et al. The application of i.v. digital subtraction angiography to cranial disease in children. *AJNR Am J Neuroradiol.* 1986;7:669–674.
112. Vatne K, Nakstad P, Lundar T. Digital subtraction angiography (DSA) in the evaluation of brain death. A comparison of conventional cerebral angiography with intravenous and intraarterial DSA. *Neuroradiology.* 1985;27:155–157.
113. Rouge M, Battistelli JM, Truong P, Marty F, Fogliani J. [Intravenous digital subtraction angiography in intensive care]. *Ann Fr Anesth Reanim.* 1985;4:85–88.
114. Lee BC, Voorhies TM, Ehrlich ME, Lipper E, Auld PA, Vannucci RC. Digital intravenous cerebral angiography in neonates. *AJNR Am J Neuroradiol.* 1984;5:281–286.
115. Gomes AS, Hallinan JM. Intravenous digital subtraction angiography in the diagnosis of brain death. *AJNR Am J Neuroradiol.* 1983;4:21–24.
116. Goodman JM, Heck LL. Confirmation of brain death at bedside by isotope angiography. *JAMA.* 1977;238:966–968.
117. Goodman JM, Mishkin FS, Dyken M. Determination of brain death by isotope angiography. *JAMA.* 1969;209:1869–1872.
118. Mishkin FS. Cerebral radionuclide angiography. *Angiology.* 1977;28:261–275.
119. Mishkin F. Determination of cerebral death by radionuclide angiography. *Radiology.* 1975;115:135–137.

120. Braunstein P, Korein J, Kricheff II, Lieberman A. Evaluation of the critical deficit of cerebral circulation using radioactive tracers (bolus technique). *Ann N Y Acad Sci.* 1978;315:143–167.

121. Ashwal S, Smith AJ, Torres F, Loken M, Chou SN. Radionuclide bolus angiography: a technique for verification of brain death in infants and children. *J Pediatr.* 1977;91:722–727.

122. Holzman BH, Curless RG, Sfakianakis GN, Ajmone-Marsan C, Montes JE. Radionuclide cerebral perfusion scintigraphy in determination of brain death in children. *Neurology.* 1983;33:1027–1031.

123. Facco E, Zucchetta P, Munari M, et al. 99mTc-HMPAO SPECT in the diagnosis of brain death. *Intensive Care Med.* 1998;24:911–917.

124. Yamamoto T, Katayama Y. Deep brain stimulation therapy for the vegetative state. *Neuropsychol Rehabil.* 2005;15:406–413.

125. Kahveci F, Bekar A, Tamgac F. Tc-99 HMPAO cerebral SPECT imaging in brain death patients with complex spinal automatism. *Ulus Travma Derg.* 2002;8:198–201.

126. Keske U. Tc-99m-HMPAO single photon emission computed tomography (SPECT) as an ancillary test in the diagnosis of brain death. *Intensive Care Med.* 1998;24:895–897.

127. Bonetti MG, Ciritella P, Valle G, Perrone E. 99mTc HM-PAO brain perfusion SPECT in brain death. *Neuroradiology.* 1995;37:365–369.

128. Spieth ME, Ansari AN, Kawada TK, Kimura RL, Siegel ME. Direct comparison of Tc-99m DTPA and Tc-99m HMPAO for evaluating brain death. *Clin Nucl Med.* 1994;19:867–872.

129. Valle G, Ciritella P, Bonetti MG, Dicembrino F, Perrone E, Perna GP. Considerations of brain death on a SPECT cerebral perfusion study. *Clin Nucl Med.* 1993;18:953–954.

130. Tsuchida T, Sadato N, Nishizawa S, et al. [99mTc-HMPAO SPECT in the brain death—a case report]. *Kaku Igaku.* 1993;30:663–667.

131. Vander BT, Laloux P, Maes A, Salmon E, Goethals I, Goldman S. Guidelines for brain radionuclide imaging. Perfusion single photon computed tomography (SPECT) using Tc-99m radiopharmaceuticals and brain metabolism positron emission tomography (PET) using F-18 fluorodeoxyglucose. The Belgian Society for Nuclear Medicine. *Acta Neurol Belg.* 2001;101:196–209.

132. Staelens L, Oltenfreiter R, Dumont F, et al. In vivo evaluation of [123I]-4–iodo-N-4-4-2-methoxyphenyl)-piperazin-1-yl)butyl)-benzamide: a potential sigma receptor ligand for SPECT studies. *Nucl Med Biol.* 2005;32:193–200.

133. Waterhouse RN. Determination of lipophilicity and its use as a predictor of blood-brain barrier penetration of molecular imaging agents. *Mol Imaging Biol.* 2003;5:376–389.

134. Weckesser M, Schober O. Brain death revisited: utility confirmed for nuclear medicine. *Eur J Nucl Med.* 1999;26:1387–1391.

135. Elsayed AA, Moran CK, Cross DT 3rd, et al. Effect of intraarterial papaverine and/or angioplasty on the cerebral veins in patients with vasospasm after subarachnoid hemorrhage due to ruptured intracranial aneurysms. *Neurosurg Focus.* 2006;21(3):E16.

136. Previgliano IJ. Assessment: transcranial Doppler ultrasonography: report of the Therapeutics and Technology Assessment Subcommittee of the American Academy of Neurology. *Neurology.* 2004;63:2457–2458.

137. Sloan MA, Alexandrov AV, Tegeler CH, et al. Assessment: transcranial Doppler ultrasonography: report of the Therapeutics and Technology Assessment Subcommittee of the American Academy of Neurology. *Neurology.* 2004;62:1468–1481.

138. Su YY, Zhao H, Zhang Y, Wang XM, Hua Y. [Studies on evaluation of brain death]. *Zhonghua Nei Ke Za Zhi.* 2004;43:250–253.

139. Dosemeci L, Dora B, Yilmaz M, Cengiz M, Balkan S, Ramazanoglu A. Utility of transcranial Doppler ultrasonography for confirmatory diagnosis of brain death: two sides of the coin. *Transplantation.* 2004;77:71–75.

140. Singh V, McCartney JP, Hemphill JC, III. Transcranial Doppler ultrasonography in the neurologic intensive care unit. *Neurol India.* 2001;49 Suppl 1:S81–S89.

141. Azevedo E, Teixeira J, Neves JC, Vaz R. Transcranial Doppler and brain death. *Transplant Proc.* 2000;32:2579–2581.

142. Kirkham FJ, Levin SD, Padayachee TS, Kyme MC, Neville BG, Gosling RG. Transcranial pulsed Doppler ultrasound findings in brain stem death. *J Neurol Neurosurg Psychiatry.* 1987;50:1504–1513.

143. McMenamin JB, Volpe JJ. Doppler ultrasonography in the determination of neonatal brain death. *Ann Neurol.* 1983;14:302–307.

144. Budingen HJ, von Reutern GM. [Noninvasive screening of cerebral death by Doppler sonography (author's translation)]. *Dtsch Med Wochenschr.* 1979;104:1347–1351.

145. Nebra AC, Virgos B, Santos S, et al. [Clinical diagnostic of brain death and transcranial Doppler, looking for middle cerebral arteries and intracranial vertebral arteries. Agreement with scintigraphic techniques]. *Rev Neurol.* 2001;33:916–920.

146. Burger R, Schlake HP, Seybold S, Reiners C, Bendszus M, Roosen K. [Value of transcranial Doppler ultrasonography compared with scintigraphic techniques and EEG in brain death]. *Zentralbl Neurochir.* 2000;61:7–13.

147. Dominguez-Roldan JM, Garcia-Alfaro C, Jimenez-Gonzalez PI, Rivera-Fernandez V, Hernandez-Hazanas F, Perez-Bernal J. Brain death due to supratentorial masses: diagnosis using transcranial Doppler sonography. *Transplant Proc.* 2004;36:2898–2900.

148. Dominguez-Roldan JM, Jimenez-Gonzalez PI, Garcia-Alfaro C, Rivera-Fernandez V, Hernandez-Hazanas F. Diagnosis of brain death by transcranial Doppler sonography: solutions for cases of difficult sonic windows. *Transplant Proc.* 2004;36:2896–2897.

149. Dominguez-Roldan JM, Murillo-Cabezas F, Munoz-Sanchez A, Santamaria-Mifsut JL, Villen-Nieto J, Barrera-Chacon JM. Study of blood flow velocities in the middle cerebral artery using transcranial Doppler sonography in brain-dead patients. *Transplant Proc.* 1995;27:2395–2396.

150. Dominguez-Roldan JM, Murillo-Cabezas F, Munoz-Sanchez A, Santamaria-Mifsut JL, Villen-Nieto J. Changes in the Doppler waveform of intracranial arteries in patients with brain-death status. *Transplant Proc.* 1995;27:2391–2392.

151. Cabrer C, Dominguez-Roldan JM, Manyalich M, et al. Persistence of intracranial diastolic flow in transcranial Doppler sonography exploration of patients in brain death. *Transplant Proc.* 2003;35:1642–1643.

152. Schoning M, Scheel P, Holzer M, Fretschner R, Will BE. Volume measurement of cerebral blood flow: assessment of cerebral circulatory arrest. *Transplantation.* 2005;80:326–331.

153. Fages E, Tembl JI, Fortea G, et al. [Clinical usefulness of transcranial Doppler in diagnosis of brain death]. *Med Clin (Barc).* 2004;122:407–412.

154. Seano-Estudillo JL, Castanon-Gonzalez JA, Carbajal-Ramirez A, Castrejon-Roman H, Leon-Gutierrez MA. [Cerebral blood flow velocity spectrum by transcranial

Doppler ultrasound in patients with brain death clinical criteria.]. *Gac Med Mex.* 2003;139:535–538.

155. de Freitas GR, Andre C. Routine insonation of the transorbital window for confirming brain death: a double-edged sword. *Arch Neurol.* 2003;60:1169.

156. Jacobs BS, Carhuapoma JR, Castellanos M. Clarifying TCD criteria for brain death—are some arteries more equal than others? *J Neurol Sci.* 2003;210:3–4.

157. Rodriguez RA, Cornel G, Alghofaili F, Hutchison J, Nathan HJ. Transcranial Doppler during suspected brain death in children: Potential limitation in patients with cardiac "shunt". *Pediatr Crit Care Med.* 2002;3:153–157.

158. Roldan DH. [Intracranial hypertension and brain death]. *Rev Esp Anestesiol Reanim.* 2002;49:225–226.

159. Sadik JC, Riquier V, Koskas P, et al. [Transcranial Doppler imaging: state of the art]. *J Radiol.* 2001;82:821–831.

160. Singh V, McCartney JP, Hemphill JC, III. Transcranial Doppler ultrasonography in the neurologic intensive care unit. *Neurol India.* 2001;49(suppl 1):S81–S89.

161. Valentin A, Karnik R, Winkler WB, Hochfellner A, Slany J. Transcranial Doppler for early identification of potential organ transplant donors. *Wien Klin Wochenschr.* 1997;109:836–839.

162. Ducrocq X, Hassler W, Moritake K, et al. Consensus opinion on diagnosis of cerebral circulatory arrest using Doppler-sonography: Task Force Group on cerebral death of the Neurosonology Research Group of the World Federation of Neurology. *J Neurol Sci.* 1998;159:145–150.

163. McDonald L. Determination of brain death via pulsatile echoencephalography. *J Neurosurg Nurs.* 1978;10:150–155.

164. Uematsu S, Walker AE. A method for recording the pulsation of the midline echo in clinical brain death. *Johns Hopkins Med J.* 1974;135:383–390.

165. Pistoia F, Johnson DW, Darby JM, Horton JA, Applegate LJ, Yonas H. The role of xenon CT measurements of cerebral blood flow in the clinical determination of brain death. *AJNR Am J Neuroradiol.* 1991;12:97–103.

166. Darby J, Yonas H, Brenner RP. Brainstem death with persistent EEG activity: evaluation by xenon-enhanced computed tomography. *Crit Care Med.* 1987;15:519–521.

167. Ashwal S, Schneider S, Thompson J. Xenon computed tomography measuring cerebral blood flow in the determination of brain death in children. *Ann Neurol.* 1989;25:539–546.

168. Thompson JR, Ashwal S, Schneider S, Hasso AN, Hinshaw DB Jr, Kirk G. Comparison of cerebral blood flow measurements by xenon computed tomography and dynamic brain scintigraphy in clinically brain dead children. *Acta Radiol Suppl.* 1986;369:675–679.

169. Latchaw RE, Yonas H, Darby JM, Gur D, Pentheny SL. Xenon/CT cerebral blood flow determination following cranial trauma. *Acta Radiol Suppl.* 1986;369:370–373.

170. Arnold H. [Brain death]. *Nervenarzt.* 1976;47:529–537.

171. Planitzer J, Zschenderlein R, Schulze HA, Lehmann R. [Computer tomography studies in irreversible cerebral function loss (brain death)]. *Psychiatr Neurol Med Psychol (Leipz).* 1985;37:509–517.

172. Shiogai T, Takeuchi K. [Evaluation of brain death by computed tomography]. *No To Shinkei.* 1983;35:1229–1239.

173. Zilkha A, Stenzler SA, Lin JH. Computed tomography of the normal and abnormal superior sagittal sinus. *Clin Radiol.* 1982;33:415–425.

174. Eick JJ, Miller KD, Bell KA, Tutton RH. Computed tomography of deep cerebral venous thrombosis in children. *Radiology.* 1981;140:399–402.

175. Arnold H, Kuhne D, Rohr W, Heller M. Contrast bolus technique with rapid CT scanning. A reliable diagnostic tool for the determination of brain death. *Neuroradiology.* 1981;22:129–132.

176. Askenazy JJ, Sazbon L, Hackett P, Najenson T. The value of electroencephalography in prolonged coma: a comparative EEG-computed axial tomography study of two patients one year after trauma. *Resuscitation.* 1980;8:181–194.

177. Schiff ND, Rodriguez-Moreno D, Kamal A, et al. fMRI reveals large-scale network activation in minimally conscious patients. *Neurology.* 2005;64:514–523.

178. Karantanas AH, Hadjigeorgiou GM, Paterakis K, Sfiras D, Komnos A. Contribution of MRI and MR angiography in early diagnosis of brain death. *Eur Radiol.* 2002;12:2710–2716.

179. Wijdicks EF, Campeau NG, Miller GM. MR imaging in comatose survivors of cardiac resuscitation. *AJNR Am J Neuroradiol.* 2001;22:1561–1565.

180. Orrison WW Jr, Champlin AM, Kesterson OL, Hartshorne MF, King JN. MR 'hot nose sign' and 'intravascular enhancement sign' in brain death. *AJNR Am J Neuroradiol.* 1994;15:913–916.

181. Jones KM, Barnes PD. MR diagnosis of brain death. *AJNR Am J Neuroradiol.* 1992;13:65–66.

182. Turner R, Le BD, Moonen CT, Despres D, Frank J. Echo-planar time course MRI of cat brain oxygenation changes. *Magn Reson Med.* 1991;22:159–166.

183. Laureys S. The neural correlate of (un)awareness: lessons from the vegetative state. *Trends Cogn Sci.* 2005;9:556–559.

184. Barnes PD, Robson CD. CT findings in hyperacute nonaccidental brain injury. *Pediatr Radiol.* 2000;30:74–81.

185. Dupas B, Gayet-Delacroix M, Villers D, Antonioli D, Veccherini MF, Soulillou JP. Diagnosis of brain death using two-phase spiral CT. *AJNR Am J Neuroradiol.* 1998;19:641–647.

186. Matsumura A, Meguro K, Tsurushima H, et al. Magnetic resonance imaging of brain death. *Neurol Med Chir (Tokyo).* 1996;36:166–171.

187. Yoo H, Kim IO, Wang KC, Cho BK. Preenhanced computed tomographic findings in brain death. *J Korean Med Sci.* 1993;8:305–307.

188. Rappaport ZH, Brinker RA, Rovit RL. Evaluation of brain death by contrast-enhanced computerized cranial tomography. *Neurosurgery.* 1978;2:230–232.

189. Aruga T, Ono K, Kawahara N, et al. [Some practical problems of clinical diagnosis of brain death]. *No To Shinkei.* 1983;35:777–785.

190. Watanabe T, Honda Y, Fujii Y, Koyama M, Tanaka R. Serial evaluation of axonal function in patients with brain death by using anisotropic diffusion-weighted magnetic resonance imaging. *J Neurosurg.* 2004;100:56–60.

191. Kumada K, Fukuda A, Yamane K, et al. [Diffusion-weighted imaging of brain death: study of apparent diffusion coefficient]. *No To Shinkei.* 2001;53:1027–1031.

192. Nakahara M, Ericson K, Bellander BM. Diffusion-weighted MR and apparent diffusion coefficient in the evaluation of severe brain injury. *Acta Radiol.* 2001;42:365–369.

193. McKinney AM, Teksam M, Felice R, et al. Diffusion-weighted imaging in the setting of diffuse cortical laminar necrosis and hypoxic-ischemic encephalopathy. *AJNR Am J Neuroradiol.* 2004;25:1659–1665.

194. Phan TG, Wijdicks EF. Diffusion-weighted magnetic resonance imaging in brain death. *Stroke*. 2000;31:1458–1459.

195. Lovblad KO, Bassetti C. Diffusion-weighted magnetic resonance imaging in brain death. *Stroke*. 2000;31:539–542.

196. Chalela JA, Kasner SE. Diffusion-weighted magnetic resonance imaging in brain death. *Stroke*. 2000;31(6):1459.

197. Mollaret P, Goulon M. Le coma dépassé (mémoire préliminaire). *Rev Neurol (Paris)*. 1959;101:3–15.

198. Schwab RS, Potts F, and Bonazzi A. EEG as an aid in determing death in the presence of cardiac activity (ethical, legal, and medical aspects). *Electroencephalogr Clin Neurophysiol*. 1963;15:147–148.

199. Rothstein TL. Recovery from near death following cerebral anoxia: A case report demonstrating superiority of median somatosensory evoked potentials over EEG in predicting a favorable outcome after cardiopulmonary resuscitation. *Resuscitation*. 2004;60:335–341.

200. Guideline three: minimum technical standards for EEG recording in suspected cerebral death. American Electroencephalographic Society. *J Clin Neurophysiol*. 1994;11:10–13.

201. Jorgensen EO. Technical contribution. Requirements for recording the EEG at high sensitivity in suspected brain death. *Electroencephalogr Clin Neurophysiol*. 1974;36:65–69.

202. Jorgensen EO. Clinical note. EEG without detectable cortical activity and cranial nerve areflexia as parameters of brain death. *Electroencephalogr Clin Neurophysiol*. 1974;36:70–75.

203. Trojaborg W, Jorgensen EO. Evoked cortical potentials in patients with "isoelectric" EEGs. *Electroencephalogr Clin Neurophysiol*. 1973;35:301–309.

204. Jorgensen EO. The EEG following circulatory and respiratory arrest. *Electroencephalogr Clin Neurophysiol*. 1971;30:273.

205. Ingvar DH. The relation between EEG, cerebral metabolism and cerebral circulation as well as their disorders caused by anoxia. *Electroencephalogr Clin Neurophysiol*. 1970;29:207.

206. Kaufer C, Penin H. The flat EEG and angiography in cerebral death. *Electroencephalogr Clin Neurophysiol*. 1970;29:210.

207. Silverman D, Masland R, Saunders MG, Schwab RS. EEG and cerebral death. The neurologist's view. *Electroencephalogr Clin Neurophysiol*. 1969;27:549.

208. Silverman D, Saunders MG, Schwab RS, Masland RL. Cerebral death and the electroencephalogram. Report of the ad hoc committee of the American Electroencephalographic Society on EEG Criteria for determination of cerebral death. *JAMA*. 1969;209:1505–1510.

209. Jonkman EJ. Cerebral death and the isoelectric EEG. *Electroencephalogr Clin Neurophysiol*. 1969;27:215.

210. Spann W. [Cerebral death in the EEG]. *Dtsch Med Wochenschr*. 1968;93:2542.

211. Pallis C. Danish ethics council rejects brain death as the criterion of death—commentary 2: return to Elsinore. *J Med Ethics*. 1990;16:10–13.

212. Pallis C. Brainstem death: the evolution of a concept. *Semin Thorac Cardiovasc Surg*. 1990;2:135–152.

213. Pallis C. Brainstem death. In: Braakman R, ed. *Handbook of Clinical Neurology: Head Injury*. Amsterdam: Elsevier Science BV, 1990:441–96.

214. Pallis C. Defining death. *Br Med J (Clin Res Ed)*. 1985;291:666–667.

215. Pallis C. ABC of brain stem death. The arguments about the EEG. *Br Med J (Clin Res Ed).* 1983;286:284–287.
216. Rodin E, Tahir S, Austin D, Andaya L. Brainstem death. *Clin Electroencephalogr.* 1985;16:63–71.
217. Guidelines for the determination of death. Report of the medical consultants on the diagnosis of death to the President's Commission for the Study of Ethical Problems in Medicine and Biomedical and Behavioral Research. *Crit Care Med.* 1982;10:62–64.
218. Bernat JL, D'Alessandro AM, Port FK, et al. Report of a national conference on donation after cardiac death. *Am J Transplant.* 2006;6:281–291.
219. Bernat JL. A defense of the whole-brain concept of death. *Hastings Cent Rep.* 1998;28:14–23.
220. Bernat JL. Brain death. Occurs only with destruction of the cerebral hemispheres and the brain stem. *Arch Neurol.* 1992;49:569–570.
221. Bernat JL. How much of the brain must die in brain death? *J Clin Ethics.* 1992;3:21–26.
222. Bernat JL. *Ethical Issues in Neurology.* Philadelphia: Lippincott, 1991.
223. Facco E, Machado C. Evoked potentials in the diagnosis of brain death. *Adv Exp Med Biol.* 2004;550:175–187.
224. A definition of irreversible coma. Report of the Ad Hoc Committee of the Harvard Medical School to Examine the Definition of Brain Death. *JAMA.* 1968;205:337–340.
225. Beecher HK. After the "definition of irreversible coma". *N Engl J Med.* 1969;281:1070–1071.
226. Buchner H, Schuchardt V. Reliability of electroencephalogram in the diagnosis of brain death. *Eur Neurol.* 1990;30:138–141.
227. Egol AB, Guntupalli KK. Intravenous infusion device artifact in the EEG-confusion in the diagnosis of electrocerebral silence. *Intensive Care Med.* 1983;9:29–32.
228. Spudis EV. Brain death. *JAMA.* 1978;239:1958–1959.
229. Ferbert A, Buchner H, Ringelstein EB, Hacke W. Isolated brain-stem death. Case report with demonstration of preserved visual evoked potentials (VEPs). *Electroencephalogr Clin Neurophysiol.* 1986;65:157–160.
230. Chatrian GE. Electrophysiologic evaluation of brain death: A critical appraisal. In: Aminoff MJ, ed. *Electrodiagnosis in Clinical Neurology.* New York: Churchill Livingstone, 1980:525–588.
231. Jorgensen EO, Trojaborg W. ["Visual evoked potentials" and diagnosis of cortical death]. *Nord Med.* 1971;86:1054–1055.
232. Trojaborg W. Artefacts in the "silent" EEG. *Electroencephalogr Clin Neurophysiol.* 1970;29:217.
233. Nuwer MR, Jordan KG. Continuous ICU EEG monitoring. *Suppl Clin Neurophysiol.* 2000;53:72–75.
234. Nuwer MR. ICU monitoring. *Electroencephalogr Clin Neurophysiol Suppl.* 1999;49:322–324.
235. Vespa PM, Nuwer MR, Juhasz C, et al. Early detection of vasospasm after acute subarachnoid hemorrhage using continuous EEG ICU monitoring. *Electroencephalogr Clin Neurophysiol.* 1997;103:607–615.
236. Hacke W. Neuromonitoring. *J Neurol.* 1985;232:125–133.
237. Nuwer MR. Continuous EEG monitoring in the intensive care unit. *Electroencephalogr Clin Neurophysiol Suppl.* 1999;50:150–155.
238. Vespa PM, Nenov V, Nuwer MR. Continuous EEG monitoring in the intensive care unit: early findings and clinical efficacy. *J Clin Neurophysiol.* 1999;16:1–13.
239. Nuwer MR. Paperless electroencephalography. *Semin Neurol.* 1990;10:178–184.

240. Nuwer MR. Quantitative electroencephalography. *Arch Gen Psychiatry.* 1987;44:840.

241. Bricolo A, Faccioli F, Grosslercher JC, Pasut ML, Pinna GP, Turazzi S. Electrophysiological monitoring in the intensive care unit. *Electroencephalogr Clin Neurophysiol Suppl.* 1987;39:255–263.

242. Bricolo A, Turazzi S, Feriotti G. Prolonged posttraumatic unconsciousness: therapeutic assets and liabilities. *J Neurosurg.* 1980;52:625–634.

243. Bricolo A, Turazzi S, Faccioli F, Odorizzi F, Sciaretta G, Erculiani P. Clinical application of compressed spectral array in long-term EEG monitoring of comatose patients. *Electroencephalogr Clin Neurophysiol.* 1978;45:211–225.

244. Schnakers C, Majerus S, Laureys S. Bispectral analysis of electroencephalogram signals during recovery from coma: preliminary findings. *Neuropsychol Rehabil.* 2005;15:381–388.

245. Escudero D, Otero J, Muniz G, et al. The Bispectral Index Scale: its use in the detection of brain death. *Transplant Proc.* 2005;37:3661–3663.

246. Vivien B, Paqueron X, Le Cosquer P, Langeron O, Coriat P, Riou B. Detection of brain death onset using the bispectral index in severely comatose patients. *Intensive Care Med.* 2002;28:419–425.

247. Kechina VV, Popova TG, Ivanov LB. [Clinical importance of compressed spectral analysis of EEG in pediatric anesthesiology]. *Anesteziol Reanimatol.* 1985;11–13.

248. Gusev EI, Fedin AI, Erokhin OY, Shvarts IP. Compressed spectral analysis of the EEG in patients with acute cerebrovascular disturbance. *Neurosci Behav Physiol.* 1985;15:144–151.

249. Gusev EI, Fedin AI, Erokhn OI, Shvarts IP. [Compressed spectral EEG analysis of patients with acute cerebral circulatory disorders]. *Zh Nevropatol Psikhiatr Im S S Korsakova.* 1981;81:1133–1141.

250. Chatrian GE, Bergamasco B, Bricolo A, Frost JD Jr, Prior PF. IFCN recommended standards for electrophysiologic monitoring in comatose and other unresponsive states. Report of an IFCN committee. *Electroencephalogr Clin Neurophysiol.* 1996;99:103–122.

251. Prior PF, Maynard DE, Brierley JB. E.E.G. monitoring for the control of anaesthesia produced by the infusion of althesin in primates. *Br J Anaesth.* 1978;50:993–1001.

252. Schwartz MS, Colvin MP, Prior PF, et al. The cerebral function monitor. Its value in predicting the neurological outcome in patients undergoing cardiopulmonary by-pass. *Anaesthesia.* 1973;28:611–618.

253. Myles PS, Cairo S. Artifact in the bispectral index in a patient with severe ischemic brain injury. *Anesth Analg.* 2004;98:706–707.

254. Barbato M. Bispectral index monitoring in unconscious palliative care patients. *J Palliat Care.* 2001;17:102–108.

255. Anez SC, Recasens UJ, Lorente CC, Bodi SM, Rull BM. [The bispectral electroencephalographic index (BIS) and brain death]. *Rev Esp Anestesiol Reanim.* 2000;47:422–423.

256. Machado C, Cuspineda E, Valdes P, et al. Assessing acute middle cerebral artery ischemic stroke by quantitative electric tomography. *Clin EEG Neurosci.* 2004;35:116–124.

257. Machado C, Valdes P, Garcia-Tigera J, et al. Brain-stem auditory evoked potentials and brain death. *Electroencephalogr Clin Neurophysiol.* 1991;80:392–398.

258. Ganes T, Lundar T. EEG and evoked potentials in comatose patients with severe brain damage. *Electroencephalogr Clin Neurophysiol.* 1988;69:6–13.

259. Schwartz G, Litscher G, Pfurtscheller G, Lechner A, Rumpl E, List WF. [Prognostic assessment and diagnosis of brain death on the intensive care unit]. *Klin Anasthesiol Intensivther.* 1994;46:262–274.

260. Litscher G, Schwarz G, Pfurtscheller G, Kleinert R, List WF. [Acoustic evoked brainstem potentials—patterns of stimulus artefacts in irreversible coma]. *EEG EMG Z Elektroenzephalogr Elektromyogr Verwandte Geb.* 1992;23:82–87.

261. Pfurtscheller G, Schwarz G, Moik H, Haase V. [Braindex—an expert system for the diagnosis of brain death]. *Biomed Tech (Berl).* 1989;34:3–8.

262. Pfurtscheller G, Schwarz G, List W. Brain death and bioelectrical brain activity. *Intensive Care Med.* 1985;11:149–153.

263. Morenski JD, Oro JJ, Tobias JD, Singh A. Determination of death by neurological criteria. *J Intensive Care Med.* 2003;18:211–221.

264. Taylor RM. Reexamining the definition and criteria of death. *Semin Neurol.* 1997; 17:265–270.

265. ten Velden GH, van Huffelen AC. [Brain death criteria; guidelines by the Public Health Council]. *Ned Tijdschr Geneeskd.* 1997;141:77–79.

266. Hirayama K. [Symptomatology and diagnostic criteria of brain death]. *No To Shinkei.* 1994;46:5–12.

267. Roosen K, Klein M. [Criteria and diagnosis of brain death]. *Versicherungsmedizin.* 1990;42:110–112.

268. Bryc S. [Criteria of brain death]. *Ann Univ Mariae Curie Sklodowska [Med].* 1988;43:1–8.

269. Kurdi A, Hijazi H. Criteria of brain death—a review. *Middle East J Anesthesiol.* 1987;9:149–161.

270. Pasnikowski W. [Current criteria of brain death]. *Przegl Lek.* 1987;44:672–675.

271. Ishiguro T. [An appraisal of the criteria of cerebral death. Review of the literature and the usefulness of single photon emission CT]. *Seishin Shinkeigaku Zasshi.* 1987;89:812–819.

272. van Alphen HA. [Brain death criteria]. *Ned Tijdschr Geneeskd.* 1983;127:2293–2294.

273. Ueki K. [Criteria for determination of brain death]. *Nippon Rinsho.* 1982;40:2516–2530.

274. Allen N, Burkholder J, Comiscioni J. Clinical criteria of brain death. *Ann N Y Acad Sci.* 1978;315:70–96.

275. Molinari GF. Review of clinical criteria of brain death. *Ann N Y Acad Sci.* 1978;315: 62–69.

276. Black PM. Criteria of brain death: review and comparison. *Postgrad Med.* 1975;57: 69–74.

277. Jacquy J, Locoge M, Mouawad E. [Criteria of brain death]. *Brux Med.* 1974;54: 385–386.

278. Miyazaki Y, Takamatsu H, Tanaka Y, Mikami N, Akagawa S. Criteria of cerebral death. *Acta Radiol Diagn (Stockh).* 1972;13:318–328.

279. Jorgensen EO, Brodersen P. [Criteria of death. The value of cliniconeurologic examination of cerebral circulation for the assessment of brain death]. *Nord Med.* 1971;86:1549–1560.

280. Guérit JM. Evoked potentials in severe brain injury. *Prog Brain Res.* 2005;150:415–426.

281. Nuwer MR, Comi G, Emerson R, et al. IFCN standards for digital recording of clinical EEG. International Federation of Clinical Neurophysiology. *Electroencephalogr Clin Neurophysiol.* 1998;106:259–261.
282. De Volder AG, Michel C, Guérit JM, et al. Brain glucose metabolism in postanoxic syndrome due to cardiac arrest. *Acta Neurol Belg.* 1994;94:183–189.
283. Guerit JM. Evoked potentials: a safe brain-death confirmatory tool? *Eur J Med.* 1992;1:233–243.
284. Kaplan PW. Electrophysiological prognostication and brain injury from cardiac arrest. *Semin Neurol.* 2006;26(4):403–412.
285. Su YY, Yang QL, Pang Y, Lv XP. Evaluation of coma patients after cardiopulmonary resuscitation. *Chin Med J (Engl).* 2005;118:1808–1811.
286. Guérit JM. [Evoked potentials and post-traumatic evolution]. *Ann Fr Anesth Reanim.* 2005;24:673–678.
287. Machado C. [Resolution for the determination and certification of death in Cuba]. *Rev Neurol.* 2003;36:763–770.
288. Tomita Y, Fukuda C, Maegaki Y, Hanaki K, Kitagawa K, Sanpei M. Re-evaluation of short latency somatosensory evoked potentials (P13, P14 and N18. for brainstem function in children who once suffered from deep coma. *Brain Dev.* 2003;25:352–356.
289. Ozgirgin ON, Ozcelik T, Sevimli NK. Auditory brain stem responses in the detection of brain death. *Kulak Burun Bogaz Ihtis Derg.* 2003;10:1–7.
290. Tamuleviciute V. [Brain death determination algorithm]. *Medicina (Kaunas).* 2002;38:792–796.
291. Ushio M, Kaga K, Sakata H, Ogawa Y, Makiyama Y, Nishimoto H. Auditory brainstem response and temporal bone pathology findings in a brain-dead infant. *Int J Pediatr Otorhinolaryngol.* 2001;58:249–253.
292. Hodelin-Tablada R, Fuentes-Pelier D. [Electroretinogram and visual evoked potentials in the diagnosis of brain death]. *Rev Neurol.* 2000;31:598–599.
293. de Tourtchaninoff M, Hantson P, Mahieu P, Guérit JM. Brain death diagnosis in misleading conditions. *QJM.* 1999;92:407–414.
294. Zwarts MJ, Kornips FH, Vogels OM. Clinical brainstem death with preserved electroencephalographic activity and visual evoked response. *Arch Neurol.* 2001;58:1010.
295. Ferbert A, Buchner H, Bruckmann H. Brainstem auditory evoked potentials and somatosensory evoked potentials in pontine haemorrhage. Correlations with clinical and CT findings. *Brain.* 1990;113(pt 1):49–63.
296. Nau HE, Wiedemayer H, Dalbah A, Engel W, Mais J. [The value of evoked potentials in the neurosurgical intensive care unit]. *Neurochirurgia (Stuttg).* 1988;31 Suppl 1:170–174.
297. Jorgensen EO. [Evoked potentials]. *Nord Med.* 1971;85:248–249.
298. Machado C, Santiesteban R, Garcia O, et al. Visual evoked potentials and electroretinography in brain-dead patients. *Doc Ophthalmol.* 1993;84:89–96.
299. Machado C, Santiesteban R, García-Tigera J, and García O. Potenciales evocados visuales y el electrorretinograma en la muerte encefálica. *Revista Cubana de Oftalmología.* 1991;4[2]:117–124.
300. Machado C. Visual evoked potentials and electroretinography in brain-dead patients. *Neurophysiol Clin.* 1990;20(suppl):18s.
301. Arfel G. [Visual stimuli and cerebral silence]. *Electroencephalogr Clin Neurophysiol.* 1967;23:172–175.

302. Wilkus RJ, Chatrian GE, Lettich E. The electroretinogram during terminal anoxia in humans. *Electroencephalogr Clin Neurophysiol.* 1971;31:537–546.

303. Walter S, Arfel G. [Responses to visual stimuli in acute and chronic coma states]. *Electroencephalogr Clin Neurophysiol.* 1972;32:27–41.

304. Reilly EL, Kondo C, Brunberg JA, Doty DB. Visual evoked potentials during hypothermia and prolonged circulatory arrest. *Electroencephalogr Clin Neurophysiol.* 1978;45:100–106.

305. Cobb WA, Dawson GD. The latency and form in man of the occipital potentials evoked by bright flashes. *J Physiol.* 1960;152:108–121.

306. Ferbert A, Buchner H, Bruckmann H, Zeumer H, Hacke W. Evoked potentials in basilar artery thrombosis: correlation with clinical and angiographic findings. *Electroencephalogr Clin Neurophysiol.* 1988;69:136–147.

307. Sament S, Alderete JF, Schwab RS. The persistence of the electroretinogram in patients with flat isoelectric EEGs. *Electroencephalogr Clin Neurophysiol.* 1969;26:121.

308. Arfel G. [Problems concerning brain death]. *Med Leg Dommage Corpor.* 1970;3:47–54.

309. Karaali K, Cevikol C, Senol U, et al. Orbital Doppler sonography findings in cases of brain death. *AJNR Am J Neuroradiol.* 2000;21:945–947.

310. Ong TJ, Harper CA, O'Day J. Restoration of ocular circulation and some visual function following external carotid endarterectomy. *Clin Exp Ophthalmol.* 2000;28:329–331.

311. Galle G, Lang GK, Ruprecht KW, Lang GE. [Importance of the orbital collateral circulation for the origin of ischemic ophthalmopathy in stenotic diseases of the internal carotid artery]. *Fortschr Neurol Psychiatr.* 1983;51:261–269.

312. Russell RW, Page NG. Critical perfusion of brain and retina. *Brain.* 1983;106(pt 2):419–434.

313. Countee RW, Vijayanathan T, Chavis P. Recurrent retinal ischemia beyond cervical carotid occlusions: clinical-angiographic correlations and therapeutic implications. *J Neurosurg.* 1981;55:532–542.

314. Chomicz J. [The usefulness of ophthalmological investigations, fluoroangioscopy, in particular, for the diagnosis of brain death: III: Fluorescein tests arm-retina and retinal circulation in brain death (author's transl)]. *Klin Oczna.* 1980;82:403–405.

315. Chomicz J. [Usefulness of ophthalmological investigations, in particular fluoroangioscopy, in the diagnosis of brain death. I. Review of literature (author's transl)]. *Klin Oczna.* 1980;82:327–329.

316. Brinck HP, Gran L, Larsen JL. [Delay of retinal fluorescence as a death criterion]. *Can Anaesth Soc J.* 1979;26:309–312.

317. Brinck HP, Gran L, Larsen JL. [Fluorescence of the retinal vessels as a criterion of ceased cerebral circulation]. *Tidsskr Nor Laegeforen.* 1979;99:364–366.

318. Sims JK. Criteria for pronouncement of death and the human brain death syndrome. *Hawaii Med J.* 1976;35:11–14.

319. Casler JA, Hoffman R, Berger L, Billinger TW, Sims JK, Bickford RG. Use of photo diode stimulation in clinical and experimental electroencephalography and electroretinography. *Electroencephalogr Clin Neurophysiol.* 1973;34:437–439.

320. Anziska BJ, Cracco RQ. Short-latency somatosensory evoked potentials to median nerve stimulation in patients with diffuse neurologic disease. *Neurology.* 1983;33:989–993.

321. Cracco RQ, Anziska BJ, Cracco JB, Vas GA, Rossini PM, Maccabee PJ. Short-latency somatosensory evoked potentials to median and peroneal nerve stimulation: studies in normal subjects and patients with neurologic disease. *Ann N Y Acad Sci.* 1982;388:412–425.

322. Anziska BJ, Cracco RQ. Short latency SEPs to median nerve stimulation: comparison of recording methods and origin of components. *Electroencephalogr Clin Neurophysiol.* 1981;52:531–539.

323. Anziska BJ, Cracco RQ. Short latency somatosensory evoked potentials in brain dead patients. *Arch Neurol.* 1980;37:222–225.

324. Anziska B, Cracco RQ, Cook AW, Feld EW. Somatosensory far field potentials: studies in normal subjects and patients with multiple sclerosis. *Electroencephalogr Clin Neurophysiol.* 1978;45:602–610.

325. Chiappa KH. Patternshift visual evoked potentials: Methodology. In: Chiappa KH, ed. *Evoked Potentials in Clinical Medicine.* New York: Raven Press, 1989:37–171.

326. Chiappa KH, Ropper AH. Evoked potentials in clinical medicine (second of two parts). *N Engl J Med.* 1982;306:1205–1211.

327. Chiappa KH, Ropper AH. Evoked potentials in clinical medicine (first of two parts). *N Engl J Med.* 1982;306:1140–1150.

328. Starr A. Auditory brain-stem responses in brain death. *Brain.* 1976;99:543–554.

329. Lumenta CB, Kramer M, von Tempelhoff W, Bock W. Brain stem auditory evoked potentials (BAEP) monitoring in brain death. *Neurosurg Rev.* 1989;12(suppl 1):317–321.

330. Goldie WD, Chiappa KH, Young RR, Brooks EB. Brainstem auditory and short-latency somatosensory evoked responses in brain death. *Neurology.* 1981;31:248–256.

331. Garcia-Larrea L, Artru F, Bertrand O, Pernier J, Mauguière F. The combined monitoring of brain stem auditory evoked potentials and intracranial pressure in coma. A study of 57 patients. *J Neurol Neurosurg Psychiatry.* 1992;55:792–798.

332. Garcia-Larrea L, Bertrand O, Artru F, Pernier J, Mauguière F. Brain-stem monitoring. II. Preterminal BAEP changes observed until brain death in deeply comatose patients. *Electroencephalogr Clin Neurophysiol.* 1987;68:446–457.

333. Hall JW, III, Tucker DA. Sensory evoked responses in the intensive care unit. *Ear Hear.* 1986;7:220–232.

334. Hall JW III, Key-Hargadine JR, Kim EE. Auditory brain-stem response in determination of brain death. *Arch Otolaryngol.* 1985;111:613–620.

335. Borel C, Hanley D. Neurologic intensive care unit monitoring. *Crit Care Clin.* 1985;1(2):223–239.

336. Firsching R. The brain-stem and 40 Hz middle latency auditory evoked potentials in brain death. *Acta Neurochir (Wien).* 1989;101(1–2):52–55.

337. Kayamori R, Orii K, Sato H. [Electrophysiological monitoring of the brainstem function in impending brain death—serial changes of blink reflex and brainstem auditory evoked potential]. *No To Shinkei.* 1989;41:337–342.

338. Chen ST. [Electroencephalography and brainstem auditory evoked potential in brain death]. *Taiwan Yi Xue Hui Za Zhi.* 1989;88:70–73.

339. Deguine C, Pulec JL. Temporal bone fracture with hemotympanum. *Ear Nose Throat J.* 2003;82:903.

340. Huff JS, Weimerskirch P. Observations on hemotympanum. *J Emerg Med.* 1989; 7:411–412.

341. Lee D, Honrado C, Har-El G, Goldsmith A. Pediatric temporal bone fractures. *Laryngoscope.* 1998;108:816–821.

342. Alvi A, Bereliani A. Trauma to the temporal bone: diagnosis and management of complications. *J Craniomaxillofac Trauma.* 1996;2:36–48.

343. Kaga K, Uebo K, Sakata H, Suzuki J. Auditory and vestibular pathology in brainstem death revealed by auditory brainstem response. *Acta Otolaryngol Suppl.* 1993;503:99–103.

344. Ghorayeb BY, Rafie JJ. [Fracture of the temporal bone. Evaluation of 123 cases]. *J Radiol.* 1989;70:703–710.

345. Kaga K, Takamori A, Mizutani T, Nagai T, Marsh RR. The auditory pathology of brain death as revealed by auditory evoked potentials. *Ann Neurol.* 1985;18:360–364.

346. Nuwer MR, Aminoff M, Goodin D, et al. IFCN recommended standards for brainstem auditory evoked potentials. Report of an IFCN committee. International Federation of Clinical Neurophysiology. *Electroencephalogr Clin Neurophysiol.* 1994;91:12–17.

347. Møller AR, Jho HD, Yokota M, Jannetta PJ. Contribution from crossed and uncrossed brainstem structures to the brainstem auditory evoked potentials: a study in humans. *Laryngoscope.* 1995;105:596–605.

348. Møller AR, Moller MB, Jannetta PJ, Jho HD. Compound action potentials recorded from the exposed eighth nerve in patients with intractable tinnitus. *Laryngoscope.* 1992;102:187–197.

349. Møller AR, Jannetta PJ. Interpretation of brainstem auditory evoked potentials: results from intracranial recordings in humans. *Scand Audiol.* 1983;12:125–133.

350. Møller AR, Jannetta PJ. Auditory evoked potentials recorded intracranially from the brain stem in man. *Exp Neurol.* 1982;78:144–157.

351. Møller AR, Jannetta P, Møller MB. Intracranially recorded auditory nerve response in man. New interpretations of BSER. *Arch Otolaryngol.* 1982;108:77–82.

352. Brodersen P, Jorgensen EO. Cerebral blood flow and oxygen uptake, and cerebrospinal fluid biochemistry in severe coma. *J Neurol Neurosurg Psychiatry.* 1974;37:384–391.

353. Machado C, García-Tigera J, and Coutin P. Multimodality evoked potentials and electroretinography in a test battery for the early diagnosis of brain death. *Electroencephalogr Clin Neurophysiol.* 1991;79(5):S19.

354. Machado C, Pumariega J, García-Tigera J, Miranda J, Coutin P, Antelo J, Hernández-Meilán O, Román J. A multimodal evoked potential and electroretinography test battery for the early diagnosis of brain death. *Int J Neurosci.* 1989;49:241–242.

355. Kim JS, Lopez I, DiPatre PL, Liu F, Ishiyama A, Baloh RW. Internal auditory artery infarction: clinicopathologic correlation. *Neurology.* 1999;52:40–44.

356. Tang A, Ura M, Yamauchi M, Noda Y. [The changes in cochlear blood flow and auditory brainstem response in the pressure-induced animal model of acoustic neuroma]. *Nippon Jibiinkoka Gakkai Kaiho.* 1996;99:370–378.

357. Tang A, Ura M, Yamauchi M, Noda Y. [Cochlear ischemia and changes in compound action potentials in the guinea pig]. *Nippon Jibiinkoka Gakkai Kaiho.* 1996;99:320–326.

358. Matsunaga T, Kanzaki J, Hosoda Y. The vasculature of the peripheral portion of the human eighth cranial nerve. *Hear Res.* 1996;101:119–131.

359. Shatari T, Hosoda Y, Kanzaki J. A study on the vasculature of the internal auditory artery in humans by casts. *Acta Otolaryngol Suppl.* 1994;514:101–107.

360. Shiogai T, Takeuchi K. Diagnostic reliability in loss of brainstem function evaluated by brainstem auditory evoked potentials and somatosensory evoked potentials

in impending brain death. In: Machado C, ed. *Brain Death (Proceedings of the Second International Symposium on Brain Death)*. Amsterdam: Elsevier Science BV, 1995:111–118.

361. Shiogai T, Takeuchi K. Multimodal neuromonitoring in impending brain death. *Minerva Anestesiol.* 1994;60:583–588.

362. Uziel A, Benezech J, Moulinier MB, Duboin MP. [Evoked brain stem potentials in cases of traumatic coma of: value in determining the degree of brain stem dysfunction (author's transl)]. *Rev Electroencephalogr Neurophysiol Clin.* 1979;9(2):202–206.

363. Uziel A, Benezech J. Auditory brain-stem responses in comatose patients: relationship with brain-stem reflexes and levels of coma. *Electroencephalogr Clin Neurophysiol.* 1978;45:515–524.

364. Uziel A, Guerrier Y. [Recording of potentials in the cochlea and brain stem via surface electrodes (author's transl)]. *Ann Otolaryngol Chir Cervicofac.* 1978;95:19–32.

365. Wytrzes LM, Chatrian GE, Shaw CM, Wirch AL. Acute failure of forebrain with sparing of brain-stem function. Electroencephalographic, multimodality evoked potential, and pathologic findings. *Arch Neurol.* 1989;46:93–97.

366. Stockard JJ, Iragui VJ. Clinically useful applications of evoked potentials in adult neurology. *J Clin Neurophysiol.* 1984;1:159–202.

367. Suzuki J, Yamane H. The choice of stimulus in the auditory brainstem response test for neurological and audiological examinations. *Ann N Y Acad Sci.* 1982;388:731–736.

368. Rowe MJ, III. The brainstem auditory evoked response in neurological disease: a review. *Ear Hear.* 1981;2:41–51.

369. Freye E. Cerebral monitoring in the operating room and the intensive care unit—an introductory for the clinician and a guide for the novice wanting to open a window to the brain. Part II: Sensory-evoked potentials (SSEP, AEP, VEP). *J Clin Monit Comput.* 2005;19:77–168.

370. Legatt AD. Mechanisms of intraoperative brainstem auditory evoked potential changes. *J Clin Neurophysiol.* 2002;19:396–408.

371. Yan Y, Tang W. Changes of evoked potentials and evaluation of mild hypothermia for treatment of severe brain injury. *Chin J Traumatol.* 2001;4:8–13.

372. Romani A, Bergamaschi R, Versino M, Zilioli A, Callieco R, Cosi V. Circadian and hypothermia-induced effects on visual and auditory evoked potentials in multiple sclerosis. *Clin Neurophysiol.* 2000;111:1602–1606.

373. Papadopoulos G, Lang M, Link J, et al. [Loss of brain stem auditory evoked potential waves I and II during controlled hypotension]. *Anaesthesist.* 1995;44:785–788.

374. Kadoi Y, Fujita N, Saito S, Fukura H, Fujita T. [Effects of anesthetic drugs and temperature on brain stem and mid-latency evoked potentials]. *Masui.* 1995;44:1213–1217.

375. Markand ON, Warren C, Mallik GS, Williams CJ. Temperature-dependent hysteresis in somatosensory and auditory evoked potentials. *Electroencephalogr Clin Neurophysiol.* 1990;77:425–435.

376. Zhou L, Shao Z, Ou S. Cryoanalgesia: electrophysiology at different temperatures. *Cryobiology.* 2003;46:26–32.

377. Wangemann P, Wonneberger K. Neurogenic regulation of cochlear blood flow occurs along the basilar artery, the anterior inferior cerebellar artery and at branch points of the spiral modiolar artery. *Hear Res.* 2005;209:91–96.

378. Grgic M, Petric V, Grgic MP, Demarin V, Pegan B. Doppler ultrasonography of the vertebrobasilar circulation in patients with sudden sensorineural hearing loss. *J Otolaryngol.* 2005;34:51–59.

379. Alekseeva NS. [Cochleovestibular syndromes in vertebrobasilar insufficiency]. *Zh Nevrol Psikhiatr Im S S Korsakova.* 2004;104:16–21.

380. Sone M, Hayashi H, Tominaga M, Nakashima T. Changes in cochlear blood flow due to endotoxin-induced otitis media. *Ann Otol Rhinol Laryngol.* 2004;113:450–454.

381. Yamamoto H, Tominaga M, Sone M, Nakashima T. Contribution of stapedial artery to blood flow in the cochlea and its surrounding bone. *Hear Res.* 2003;186:69–74.

382. Zegers de Beyl D, Brunko E. Thoughts about the IFCN recommendations for the recording of short latency somatosensory evoked potentials. *Electroencephalogr Clin Neurophysiol.* 1995;94:151–153.

383. Brunko E, Delecluse F, Herbaut AG, Levivier M, Zegers de Beyl D. Unusual pattern of somatosensory and brain-stem auditory evoked potentials after cardio-respiratory arrest. *Electroencephalogr Clin Neurophysiol.* 1985;62:338–342.

384. [Clinical use of transcranial Doppler in critical neurological patients. Results of a multicenter study]. *Med Clin (Barc).* 2003;120:241–245.

385. Flowers WM Jr, Patel BR. Accuracy of clinical evaluation in the determination of brain death. *South Med J.* 2000;93:203–206.

386. Jarus-Dziedzic K, Zub W, Czernicki Z. [The application of transcranial Doppler sonography in the evaluation of cerebral flow disturbances after head trauma]. *Neurol Neurochir Pol.* 1999;33:151–167.

387. Kawamoto M, Sera A, Kaneko K, Yuge O, Ohtani M. Parasympathetic activity in brain death: effect of apnea on heart rate variability. *Acta Anaesthesiol Scand.* 1998;42:47–51.

388. Leung H, Kwan P, Elger CE. Finding the missing link between ictal bradyarrhythmia, ictal asystole, and sudden unexpected death in epilepsy. *Epilepsy Behav.* 2006;9(1):19–30.

389. Su CF, Kuo TB, Kuo JS, Lai HY, Chen HI. Sympathetic and parasympathetic activities evaluated by heart-rate variability in head injury of various severities. *Clin Neurophysiol.* 2005;116:1273–1279.

390. Madanmohan, Udupa K, Bhavanani AB, Shatapathy CC, Sahai A. Modulation of cardiovascular response to exercise by yoga training. *Indian J Physiol Pharmacol.* 2004;48:461–465.

391. Telles S, Joshi M, Dash M, Raghuraj P, Naveen KV, Nagendra HR. An evaluation of the ability to voluntarily reduce the heart rate after a month of yoga practice. *Integr Physiol Behav Sci.* 2004;39:119–125.

392. Luckraz H, Charman SC, Parameshwar J, et al. Are non-brain stem-dead cardiac donors acceptable donors? *J Heart Lung Transplant.* 2004;23:330–333.

393. Biswas AK, Summerauer JF. Heart rate variability and brain death. *J Neurosurg Anesthesiol.* 2004;16:62.

394. Baillard C, Vivien B, Mansier P, et al. Brain death assessment using instant spectral analysis of heart rate variability. *Crit Care Med.* 2002;30:306–310.

395. Conci F, Di RM, Castiglioni P. Blood pressure and heart rate variability and baroreflex sensitivity before and after brain death. *J Neurol Neurosurg Psychiatry.* 2001;71:621–631.

396. Neumann T, Post H, Ganz RE, et al. Linear and non-linear dynamics of heart rate variability in brain dead organ donors. *Z Kardiol.* 2001;90:484–491.

397. Baillard C, Goncalves P, Mangin L, Swynghedauw B, Mansier P. Use of time frequency analysis to follow transitory modulation of the cardiac autonomic system in clinical studies. *Auton Neurosci.* 2001;90:24–28.

398. Rapenne T, Moreau D, Lenfant F, et al. Could heart rate variability predict outcome in patients with severe head injury? A pilot study. *J Neurosurg Anesthesiol.* 2001;13:260–268.

399. Huttemann E, Schelenz C, Sakka SG, Reinhart K. Atropine test and circulatory arrest in the fossa posterior assessed by transcranial Doppler. *Intensive Care Med.* 2000;26:422–425.

400. Biswas AK, Scott WA, Sommerauer JF, Luckett PM. Heart rate variability after acute traumatic brain injury in children. *Crit Care Med.* 2000;28:3907–3912.

401. Freitas J, Azevedo E, Teixeira J, Carvalho MJ, Costa O, Falcao de FA. Heart rate variability as an assessment of brain death. *Transplant Proc.* 2000;32:2584–2585.

402. Rapenne T, Moreau D, Lenfant F, Boggio V, Cottin Y, Freysz M. Could heart rate variability analysis become an early predictor of imminent brain death? A pilot study. *Anesth Analg.* 2000;91:329–336.

403. Sinex DG, Lopez DE, Warr WB. Electrophysiological responses of cochlear root neurons. *Hear Res.* 2001;158:28–38.

404. Warr WB, Boche JB, Neely ST. Efferent innervation of the inner hair cell region: origins and terminations of two lateral olivocochlear systems. *Hear Res.* 1997;108:89–111.

405. Warr WB. Efferent components of the auditory system. *Ann Otol Rhinol Laryngol Suppl.* 1980;89:114–120.

406. Pisani R, De MM, Arbinolo MA, et al. [Contribution of Doppler ultrasound flowmetry in the evaluation of intracranial impedance. II. Monitoring intracranial resistance as a guide in therapeutic treatment]. *Minerva Anestesiol.* 1985;51:387–392.

407. Liboni M, Bertino R, Franco C, et al. [Contribution of Doppler ultrasound flowmetry in the evaluation of intracranial impedance. I. Behavior of the ultrasonic information at the carotid level in cerebral coma]. *Minerva Anestesiol.* 1985;51:381–385.

408. Semeniutin VB, Eremeev VS, Lonskii AV, Orlov VA. [Analysis of the possibility of increasing the intracranial impedance component of the rheoencephalogram using screened electrodes]. *Med Tekh.* 1981;4:29–32.

409. Vainshtein GB, Iripkhanov BB. [Changes in intracranial impedance during cerebral dehydration]. *Fiziol Zh SSSR Im I M Sechenova.* 1980;66:777–782.

410. Moller AR, Jannetta PJ, Sekhar LN. Contributions from the auditory nerve to the brain-stem auditory evoked potentials (BAEPs): results of intracranial recording in man. *Electroencephalogr Clin Neurophysiol.* 1988;71:198–211.

411. Scherg M, von Cramon D. A new interpretation of the generators of BAEP waves I-V: results of a spatio-temporal dipole model. *Electroencephalogr Clin Neurophysiol.* 1985;62:290–299.

412. Buchner H, Ferbert A, Bruckmann H, Zeumer H, Hacke W. [Validity of early acoustically-evoked potentials in the diagnosis of brain death]. *EEG EMG Z Elektroenzephalogr Elektromyogr Verwandte Geb.* 1986;17:117–122.

413. Uziel A, Benezech J, Lorenzo S, Monstrey Y, Duboin MP, Roquefeuil B. Clinical applications of brainstem auditory evoked potentials in comatose patients. *Adv Neurol.* 1982;32:195–202.
414. Biniek R, Ferbert A, Buchner H, Bruckmann H. Loss of brainstem acoustic evoked potentials with spontaneous breathing in a patient with supratentorial lesion. *Eur Neurol.* 1990;30:38–41.
415. Fauvage B, Combes P. Isoelectric electroencephalogram and loss of evoked potentials in a patient who survived cardiac arrest. *Crit Care Med.* 1993;21:472–475.
416. Facco E, Munari M, Gallo F, et al. Role of short latency evoked potentials in the diagnosis of brain death. *Clin Neurophysiol.* 2002;113:1855–1866.
417. Barelli A, Della CF, Calimici R, Sandroni C, Proietti R, Magalini SI. Do brainstem auditory evoked potentials detect the actual cessation of cerebral functions in brain dead patients? *Crit Care Med.* 1990;18:322–323.
418. Guérit JM, Tourtchaninoff M De, Hantson P, Mahieu P. Miltimodality evoked potentials in the differential diagnosis of brain death. In: Machado C, ed. *Brain Death (Proceedings of the Second International Symposium on Brain Death).* Amsterdam: Elsevier Science BV, 1995:119–26.
419. Chandler JM, Brilli RJ. Brainstem encephalitis imitating brain death. *Crit Care Med.* 1991;19:977–979.
420. Hantson P, Guérit JM, De Tourtchaninoff M. et al. Rabies encephalitis mimicking the electrophysiological pattern of brain death. A case report. *Eur Neurol.* 1993;33:212–217.
421. Loiselle DL, Nuwer MR. When should we warn the surgeon? Diagnosis-based warning criteria for BAEP monitoring. *Neurology.* 2005;65:1522–1523.
422. Chiappa KH, Ropper AH. Evoked potentials in clinical medicine (first of two parts). *N Engl J Med.* 1982;306:1140–1150, May 13, 1982.
423. Aminoff MJ. The use of somatosensory evoked potentials in the evaluation of the central nervous system. *Neurol Clin.* 1988;6:809–823.
424. Aminoff MJ. The clinical role of somatosensory evoked potential studies: a critical appraisal. *Muscle Nerve.* 1984;7:345–354.
425. Belsh JM, Chokroverty S. Short-latency somatosensory evoked potentials in brain-dead patients. *Electroencephalogr Clin Neurophysiol.* 1987;68:75–78.
426. Besser R, Dillmann U, Henn M. Somatosensory evoked potentials aiding the diagnosis of brain death. *Neurosurg Rev.* 1988;11:171–175.
427. Brunko E, Zegers de Beyl D. Prognostic value of early cortical somatosensory evoked potentials after resuscitation from cardiac arrest. *Electroencephalogr Clin Neurophysiol.* 1987;66:15–24.
428. Buchner H, Schildknecht M, Ferbert A. [Spinal and subcortical somatosensory evoked potentials: a comparison with the localization of spinal, medullary and pontine lesions and in brain death]. *EEG EMG Z Elektroenzephalogr Elektromyogr Verwandte Geb.* 1991;22:51–61.
429. Buchner H, Ferbert A, Hacke W. Serial recording of median nerve stimulated subcortical somatosensory evoked potentials (SEPs) in developing brain death. *Electroencephalogr Clin Neurophysiol.* 1988;69:14–23.
430. Butinar D, Gostisa A. Brainstem auditory evoked potentials and somatosensory evoked potentials in prediction of posttraumatic coma in children. *Pflugers Arch.* 1996;431:R289–R290.

431. Chancellor AM, Frith RW, Shaw NA. Somatosensory evoked potentials following severe head injury: loss of the thalamic potential with brain death. *J Neurol Sci.* 1988;87:255–263.

432. Cohen SN, Potvin A, Syndulko K, Pettler-Jennings P, Potvin JH, Tourtellotte WW. Multimodality evoked potentials: clinical applications and assessment of utility. *Bull Los Angeles Neurol Soc.* 1982;47:55–61.

433. de la Torre JC. Evaluation of brain death using somatosensory evoked potentials. *Biol Psychiatry.* 1981;16:931–935.

434. De Meirleir LJ, Taylor MJ. Prognostic utility of SEPs in comatose children. *Pediatr Neurol.* 1987;3:78–82.

435. Delberghe X, Brunko E, Mavroudakis N, Zegers de BD. Somatosensory central conduction time after sural and tibial nerve stimulation. *Acta Neurol Belg.* 1994;94:251–255.

436. Delberghe X, Mavroudakis N, Zegers de BD, Brunko E. The effect of stimulus frequency on post- and pre-central short-latency somatosensory evoked potentials (SEPs). *Electroencephalogr Clin Neurophysiol.* 1990;77:86–92.

437. Despland PA, Regli F. [Early evoked auditory potentials for the assessment of prognosis in cerebral resuscitation]. *Schweiz Med Wochenschr.* 1982;112:997–1001.

438. Despland PA, Regli F. [Significance of early auditory and somatesthetic evoked potentials in coma]. *Rev Med Suisse Romande.* 1985;105:323–329.

439. Facco E, Casartelli LM, Munari M, Toffoletto F, Baratto F, Giron GP. Short latency evoked potentials: new criteria for brain death? *J Neurol Neurosurg Psychiatry.* 1990;53:351–353.

440. Facco E, Caputo P, Casartelli LM, et al. Auditory and somatosensory evoked potentials in brain-dead patients. *Riv Neurol.* 1988;58:140–145.

441. Fotiou F, Tsitsopoulos P, Sitzoglou K, Fekas L, Tavridis G. Evaluation of the somatosensory evoked potentials in brain death. *Electromyogr Clin Neurophysiol.* 1987;27:55–60.

442. Haupt WF. [Multimodality evoked potentials and brain death. Prerequisites, findings and problems]. *Nervenarzt.* 1987;58:653–657.

443. Haupt WF. Evoked potentials in non-traumatic brain death. *Neurosurg Rev.* 1989;12(suppl 1):369–374.

444. Hilz MJ, Litscher G, Weis M, et al. Continuous multivariable monitoring in neurological intensive care patients—preliminary reports on four cases. *Intensive Care Med.* 1991;17:87–93.

445. key-Hargadine JR, Hall JW III. Sensory evoked responses in head injury. *Cent Nerv Syst Trauma.* 1985;2:187–206.

446. Lutschg J, Pfenninger J, Ludin HP, Vassella F. Brain-stem auditory evoked potentials and early somatosensory evoked potentials in neurointensively treated comatose children. *Am J Dis Child.* 1983;137:421–426.

447. Mauguière F, Grand C, Fischer C, Courjon J. [Aspects of early somatosensory and auditory evoked potentials in neurologic comas and brain death]. *Rev Electroencephalogr Neurophysiol Clin.* 1982;12:280–285.

448. Mavroudakis N, Brunko E, Delberghe X, Zegers de Beyl D. Dissociation of P13–P14 far-field potentials: clinical and MRI correlation. *Electroencephalogr Clin Neurophysiol.* 1993;88:240–242.

449. More RC, Nuwer MR, Dawson EG. Cortical evoked potential monitoring during spinal surgery: sensitivity, specificity, reliability, and criteria for alarm. *J Spinal Disord.* 1988;1:75–80.

450. Nau HE, Wiedemayer H, Brune-Nau R, Pohlen G, Kilian F. [Validity of the electroen-cephalogram and evoked potentials in the diagnosis of brain death]. *Anasth Intensivther Notfallmed.* 1987;22:273–277.
451. Wagner W. Scalp, earlobe and nasopharyngeal recordings of the median nerve somatosensory evoked P14 potential in coma and brain death. Detailed latency and amplitude analysis in 181 patients. *Brain.* 1996;119(pt 5):1507–1521.
452. Wagner W. [The value of differential recording of subcortical somatosensory evoked potentials (SEP) in neurosurgery]. *Zentralbl Neurochir.* 1996;57:89–96.
453. Wagner W, Ungersbock K, Perneczky A. Preserved cortical somatosensory evoked potentials in apnoeic coma with loss of brain-stem reflexes: case report. *J Neurol.* 1993;240:243–248.
454. Sonoo M, Mochizuki A, Fukuda H, et al. Lower cervical origin of the P13–like potential in median SSEPS. *J Clin Neurophysiol.* 2001;18:185–190.
455. Sonoo M. Anatomic origin and clinical application of the widespread N18 potential in median nerve somatosensory evoked potentials. *J Clin Neurophysiol.* 2000;17:258–268.
456. Sonoo M, Tsai-Shozawa Y, Aoki M, et al. N18 in median somatosensory evoked potentials: a new indicator of medullary function useful for the diagnosis of brain death. *J Neurol Neurosurg Psychiatry.* 1999;67:374–378.
457. Nuwer MR, Aminoff M, Desmedt J, et al. IFCN recommended standards for short latency somatosensory evoked potentials. Report of an IFCN committee. International Federation of Clinical Neurophysiology. *Electroencephalogr Clin Neurophysiol.* 1994;91:6–11.
458. Desmedt JE, Cheron G. Non-cephalic reference recording of early somatosensory potentials to finger stimulation in adult or aging normal man: differentiation of widespread N18 and contralateral N20 from the prerolandic P22 and N30 components. *Electroencephalogr Clin Neurophysiol.* 1981;52:553–570.
459. Mauguière F, Desmedt JE, Courjon J. Neural generators of N18 and P14 far-field somatosensory evoked potentials studied in patients with lesion of thalamus or thalamo-cortical radiations. *Electroencephalogr Clin Neurophysiol.* 1983;56:283–292.
460. Urasaki E, Wada S, Kadoya C, Yokota A, Matsuoka S, Shima F. Origin of scalp far-field N18 of SSEPs in response to median nerve stimulation. *Electroencephalogr Clin Neurophysiol.* 1990;77:39–51.
461. Tomberg C, Desmedt JE, Ozaki I, Noel P. Nasopharyngeal recordings of somatosensory evoked potentials document the medullary origin of the N18 far-field. *Electroencephalogr Clin Neurophysiol.* 1991;80:496–503.
462. Wagner W. SEP testing in deeply comatose and brain dead patients: the role of nasopharyngeal, scalp and earlobe derivations in recording the P14 potential. *Electroencephalogr Clin Neurophysiol.* 1991;80:352–363.
463. Møller AR, Jannetta PJ, Burgess JE. Neural generators of the somatosensory evoked potentials: recording from the cuneate nucleus in man and monkeys. *Electroencephalogr Clin Neurophysiol.* 1986;65:241–248.
464. Sonoo M, Hagiwara H, Motoyoshi Y, Shimizu T. Preserved widespread N18 and progressive loss of P13/14 of median nerve SEPs in a patient with unilateral medial medullary syndrome. *Electroencephalogr Clin Neurophysiol.* 1996;100:488–492.
465. Sonoo M, Sakuta M, Shimpo T, Genba K, Mannen T. Widespread N18 in median nerve SEP is preserved in a pontine lesion. *Electroencephalogr Clin Neurophysiol.* 1991;80:238–240.

466. Facco E, Munari M, Baratto F, Dona B, Ori C, Giron GP. [Evoked potentials in the diagnosis of brain death]. *Minerva Anestesiol.* 1993;59:71–74.
467. Tomita Y, Fukuda C, Maegaki Y, Hanaki K, Kitagawa K, Sanpei M. Re-evaluation of short latency somatosensory evoked potentials (P13, P14 and N18. for brainstem function in children who once suffered from deep coma. *Brain Dev.* 2003;25:352–356.
468. Roncucci P, Lepori P, Mok MS, Bayat A, Logi F, Marino A. Nasopharyngeal electrode recording of somatosensory evoked potentials as an indicator in brain death. *Anaesth Intensive Care.* 1999;27:20–25.
469. Fukuda M, Kameyama S, Honda Y, et al. Short-latency somatosensory evoked potentials in patients with brain stem tumor: study of N20 and N18 potentials. *Neurol Med Chir (Tokyo).* 1997;37:525–531.
470. Noel P, Ozaki I, Desmedt JE. Origin of N18 and P14 far-fields of median nerve somatosensory evoked potentials studied in patients with a brain-stem lesion. *Electroencephalogr Clin Neurophysiol.* 1996;98:167–170.
471. Machado C. Evoked potentials in brain death. *Clin Neurophysiol.* 2004;115:238–239.
472. Maccabee PJ, Pinkhasov EI, Cracco RQ. Short latency somatosensory evoked potentials to median nerve stimulation: effect of low frequency filter. *Electroencephalogr Clin Neurophysiol.* 1983;55:34–44.
473. Machado C. Early diagnosis of brain death. *Clin Intensive Care (Suppl).* 1992;3(2):116.
474. Machado C, Wagner A, Coutin P, et al. Potenciales Evocados Somatosensoriales de corta latencia. II- Tiempo de Conducción Central. *Rev Hosp Psiquiátr Habana.* 1988;211–221.
475. Facco E, Munari M, Baratto F, Behr AU, Giron GP. Multimodality evoked potentials (auditory, somatosensory and motor) in coma. *Neurophysiol Clin.* 1993;23:237–258.
476. Machado C, García-Tigera J, García O, García-Pumariega J, Román J. Muerte Encefálica. Criterios diagnósticos. *Rev Cubana Med.* 1991;30(3):181–206.
477. Ferbert A, Buchner H, Ringelstein EB, Hacke W. Brain death from infratentorial lesions: clinical neurophysiological and transcranial Doppler ultrasound findings. *Neurosurg Rev.* 1989;12(suppl 1):340–347.
478. Machado C, Román JM, García-Tigera J, García O, and Miranda J. Utilidad de los potenciales auditivos de tallo encefálico y somato- sensoriales corta latencia en el neuromonitoreo. *Acta Médica (Hospital Hermanos Ameijeiras).* 1990;1:95–108.
479. Heart rate variability. Standards of measurement, physiological interpretation, and clinical use. Task Force of the European Society of Cardiology and the North American Society of Pacing and Electrophysiology. *Eur Heart J.* 1996;17:354–381.
480. Heart rate variability: standards of measurement, physiological interpretation and clinical use. Task Force of the European Society of Cardiology and the North American Society of Pacing and Electrophysiology. *Circulation.* 1996;93:1043–1065.
481. Garcia OD, Machado C, Román JM, et al. Heart rate variability in coma and brain death. In: Machado C, ed. *Brain Death (Proceedings of the Second International Symposium on Brain Death).* Amsterdam: Elsevier Science BV, 1995.
482. Knuttgen D, zur Nieden K, Muller-Gorges MR, Jahn M, Doehn M. Computer aided analysis of heart rate variability in brain death. *Int J Clin Monit Comput.* 1997;14:37–42.
483. Campbell HA, Klepacki JZ, Egginton S. A new method in applying power spectral statistics to examine cardio-respiratory interactions in fish. *J Theor Biol.* 2006;21:241(2):410–419.

484. Takalo R, Hytti H, Ihalainen H. Tutorial on univariate autoregressive spectral analysis. *J Clin Monit Comput.* 2005;19:401–410.

485. Correia D, Rodrigues De Resende LA, Molina RJ, et al. Power spectral analysis of heart rate variability in HIV-infected and AIDS patients. *Pacing Clin Electrophysiol.* 2006;29:53–58.

486. Pichon A, Roulaud M, Antoine-Jonville S, de Bisschop C, Denjean A. Spectral analysis of heart rate variability: interchangeability between autoregressive analysis and fast Fourier transform. *J Electrocardiol.* 2006;39:31–37.

487. Gaul-Alacova P, Boucek J, Stejskal P, Kryl M, Pastucha P, Pavlik F. Assessment of the influence of exercise on heart rate variability in anxiety patients. *Neuro Endocrinol Lett.* 2005;26:713–718.

488. Busek P, Vankova J, Opavsky J, Salinger J, Nevsimalova S. [Heart rate variability during sleep]. *Cas Lek Cesk.* 2005;144:685–688.

489. Gujjar AR, Sathyaprabha TN, Nagaraja D, Thennarasu K, Pradhan N. Heart rate variability and outcome in acute severe stroke: role of power spectral analysis. *Neurocrit Care.* 2004;1:347–353.

490. Jokkel G, Bonyhay I, Kollai M. Heart rate variability after complete autonomic blockade in man. *J Auton Nerv Syst.* 1995;51:85–89.

491. Kollai M, Jokkel G, Bonyhay I, Tomcsanyi J, Naszlady A. Relation between tonic sympathetic and vagal control of human sinus node function. *J Auton Nerv Syst.* 1994;46:273–280.

492. Lombardi F, Malliani A, Pagani M, Cerutti S. Heart rate variability and its sympathovagal modulation. *Cardiovasc Res.* 1996;32:208–216.

493. Malliani A, Pagani M, Lombardi F. Methodological aspects of noninvasive analysis of autonomic regulation of cardiovascular variability. *Clin Sci (Lond).* 1996;91(suppl):68–71.

494. Malliani A, Pagani M, Lombardi F. Power spectral analysis of heart rate variability and baroreflex gain. *Clin Sci (Lond).* 1995;89:555–556.

495. Montano N, Ruscone TG, Porta A, Lombardi F, Pagani M, Malliani A. Power spectrum analysis of heart rate variability to assess the changes in sympathovagal balance during graded orthostatic tilt. *Circulation.* 1994;90:1826–1831.

496. Malliani A, Pagani M, Lombardi F. Importance of appropriate spectral methodology to assess heart rate variability in the frequency domain. *Hypertension.* 1994;24:140–142.

497. Malliani A, Pagani M, Lombardi F. Physiology and clinical implications of variability of cardiovascular parameters with focus on heart rate and blood pressure. *Am J Cardiol.* 1994;73:3C-9C.

498. Mayorov o, Baevsky RM. Application of space technologies for valuation of a stress level. *Stud Health Technol Inform.* 1999;68:352–356.

499. Schwarz G, Pfurtscheller G, Litscher G, List WF. Quantification of autonomic activity in the brainstem in normal, comatose and brain dead subjects using heart rate variability. *Funct Neurol.* 1987;2:149–154.

500. Freitas J, Puig J, Rocha AP, et al. Heart rate variability in brain death. *Clin Auton Res.* 1996;6:141–146.

501. La Rovere MT, Schwartz PJ. Baroreflex sensitivity as a cardiac and arrhythmia mortality risk stratifier. *Pacing Clin Electrophysiol.* 1997;20:2602–2613.

502. La Rovere MT, Mortara A, Specchia G, Schwartz PJ. Myocardial infarction and baroreflex sensitivity. Clinical studies. *G Ital Cardiol.* 1992;22:639–645.

503. Muhlnickel B. [The value of heart frequency variability in the prognostic evaluation of patients with severe cerebral injuries. 1. Experimental clinical studies]. *Anaesthesiol Reanim.* 1990;15:342–350.

504. Muhlnickel B, Brandt W. [Effect of controlled ventilation on heart rate variability in patients with severe cerebral damage]. *Psychiatr Neurol Med Psychol Beih.* 1990;43:130–134.

505. Muhlnickel B. [Variability of heart rate as a diagnostic criterion for determining brain death]. *Anaesthesiol Reanim.* 1989;14:71–81.

506. Pfurtscheller G, Druschky K, Kamp HD, et al. [Multimodal evoked potentials and heart rate variability in comatose patients—1: Method of measuring and normal values]. *EEG EMG Z Elektroenzephalogr Elektromyogr Verwandte Geb.* 1987;18:108–114.

507. Fortrat JO, Yamamoto Y, Hughson RL. Respiratory influences on non-linear dynamics of heart rate variability in humans. *Biol Cybern.* 1997;77:1–10.

508. Sugihara G, Allan W, Sobel D, Allan KD. Nonlinear control of heart rate variability in human infants. *Proc Natl Acad Sci U S A.* 1996;93:2608–2613.

6
Brain Death in Children

Although a brain death (BD) diagnosis in children encloses the same concept and similar diagnostic procedure as in adults, it is a more difficult task for physicians and nurses because of several ethical and psycho-emotional issues. Parents and the general population are more reluctant to accept death in children, and find it difficult to understand an explanation that a child with preserved heartbeats but no brain activity is dead.[1,2]

Moreover, according to Ashwal,[3] the clinical exam of suspected brain-dead pediatric patients is more difficult to carry out because of the smaller size of the patients, the immaturity of the reflexes that change with central nervous system (CNS) maturation, and the existence of fontanels and open sutures in neonates and infants.

The most frequent etiology of BD in children is trauma, due to motor vehicle accidents, shaken baby syndrome, and abuse. Asphyxia producing anoxia is also a common cause, in home accidents such as near drowning, suffocation, and from the so-called sudden infant death syndrome (SIDS). Some massive infections of the brain can also be the origin of permanent destruction of CNS structures.[3–5] The Cuban Commission[1,6] has accepted with some modifications the guidelines of the U.S. Task Force (Table 6.1), published in 1987, that has been also widely introduced in pediatric hospitals.[7] The commission has also proposed the use of ancillary tests to shorten the period of observation in children older than 1 year.[1,6]

Mejia and Pollack[8] developed a prospective cohort study for selecting children undergoing BD evaluations from a total of 5415 consecutive intensive care unit (ICU) admissions. They found substantial variability among the criteria used by clinicians for the diagnosis of BD, and some practices were contradictory to the Guidelines for the Determination of Brain Death in Children and to apnea testing methodology.

Compared with BD diagnostic criteria in adults, these guidelines do not differ in any important way. The main differences are the age-related periods of observation and the requirement of particular ancillary tests in children under 1 year of age. As in adults, in children over 1 year of age the diagnosis is based on clinical examination, and confirmatory tests are not mandatory.[9–14] Ashwal has proposed BD criteria for term infants younger than 7 days of age.[4,11,15]

TABLE 6.1. Guidelines for brain death determination in children

A. History: determine the cause of coma to eliminate reversible conditions
B. Physical examination criteria:
  1. Coma and apnea
  2. Absence of brainstem function
    a. Midposition or fully dilated pupils
    b. Absence of spontaneous oculocephalic (doll's eye) and caloric-induced eye movements
    c. Absence of movement of bulbar musculature, corneal, gag, cough, sucking, and rooting
      reflexes
    d. Absence of respiratory effort with standardized testing for apnea
  3. Patient must not be hypothermic or hypotensive
  4. Flaccid tone and absence of spontaneous or induced movements, excluding activity mediated
    at spinal cord level
  5. Examination should remain consistent for brain death throughout the predetermined period
    of observation
C. Observation period according to age:
  1. 7 days to 2 months: Two examinations and electroencephalograms (EEGs) 48 hours apart
  2. 2 months to 1 year: Two examinations and EEGs 24 hours apart or one examination and an
    initial EEG showing electrocerebral silence (ECS) combined with a radionuclide angiogram
    showing no cerebral blood flow (CBF), or both
  3. More than 1 year: Two examinations 12 to 24 hours apart; EEG and isotope angiography are
    optional; multimodality evoked potentials and Doppler ultrasonography (TCD) can be used
    as ancillary tests to shorten periods of observation

Note: Modified from the Guidelines of the Task Force for the Determination of Brain Death in Children.[7]

One of the main differences with adult BD protocols is the exploration of brainstem reflexes, because the maturation of the nervous system entails different periods of time for a specific reflex to be fully developed. The pupil light reflex is only present after 30 weeks' gestation and the oculocephalic response is absent before 32 weeks. Moreover, as has already been discussed, the small size of the infant's head and body imposes added difficulties in assessing brainstem reflexes. Other complications in the neonatology critical care unit are related to the placement of the intubated infants in incubators, with adhesive tape around the face and cheek to restrict movements and protect the tracheal tube. These conditions make it difficult to perform the neurological exam.[3,5,8,11–14,16–25]

In addition, Ashwal[3] has detailed other factors that make it difficult to perform the neurological examination in infants:

• Newborn's pupils are small and barely pigmented, making it difficult to assess the light reflex. Moreover, corneal injury, retinal hemorrhages, and eyelid swelling impair the assessment of pupil responses.
• Regarding ocular motility, the exploration of oculovestibular responses by caloric stimulation is difficult because of the small size of the external ear canals. Hence, it is recommended to explore both the oculocephalic and the oculovestibular (caloric) responses.
• Regarding the corneal reflex and dehydration, irritation, and maceration of the cornea, the use of eye patches hampers this response.

- Intubation and the use of adhesive tape around the face and cheek affect testing the gag and cough reflex. These reflexes are better explored by suctioning with a catheter through the endotracheal tube.

Apnea Test

The technique for this test in children is the same as in adults. The threshold for apnea for CO_2 for children has been assumed to be the same as that for adults[8,9,24,26–28]: a pCO_2 of 60 torr. Normalization of carbon dioxide tension and reoxygenation before beginning the apnea test is recommended as well as careful monitoring of the heart rate and blood pressure during the procedure. Prolonged bradycardia or the development of hypotension during testing is likely due to irreversible brainstem failure.

Vardis and Pollack[29] reported a 4-year-old boy who fulfilled all the criteria for BD except the apnea test. An apnea test was performed for 9 minutes and 23 seconds, at which time spontaneous respiratory effort was initiated with a pH of 7.08 and a $PaCO_2$ of 91 torr (12.1 kPa). These authors concluded that the current guidelines for apnea testing may lead to erroneous evaluation of the medullary-respiratory drive. Nonetheless, according to Lang,[30–35] an examiner with experience can apply this test in 99.9% of all cases; the remaining ones may be studied using other ancillary tests.

Brain Death Determination in the Newborn

Brain death in the newborn can be diagnosed provided the physician is aware of the limitations of the clinical examination and laboratory testing. It is important to carefully and repeatedly examine these infants, with particular attention to the examination of brainstem reflexes and apnea testing. Brain death can be determined in preterm and term infants but usually requires reliance on confirmatory neurodiagnostic testing.[4,5,11,17]

As the newborn has patent sutures and open fontanels, the usual cascade of events of herniation due to an increased intracranial pressure (ICP) and reduced cerebral perfusion are less probable to occur in the newborn. Some reports have considered the effects of immaturity of the nervous system and the consequences of development on the pathophysiology and diagnosis of BD in the preterm and term infant. In the neonate, even though the cerebral blood flow (CBF) and mean arterial blood pressures are much lower, increases in ICP after acute injury are less dramatic.[3,4,5,11]

Ashwal[4] estimated that each year about 550 newborns out of a total of 4,900,000 live births might be diagnosed as brain dead. He also commented that most authors believe that the ability to diagnose BD in newborns is uncertain, primarily due to the small number of brain-dead neonates reported in the literature.[11] This explains why preterm and term neonates less than 7 days of age were excluded from the 1987 Task Force Guidelines.[23] Ashwal published data from 18 brain-dead neonates and suggested that the BD diagnosis could be achieved in term infants and preterm infants older than 34 weeks gestational age within the first week of life.[36]

Regarding confirmatory tests, a higher incidence of brain-dead newborns with preservation of electroencephalogram (EEG) activity or cerebral perfusion has been reported. Some newborns with isoelectric EEGs showed preserved CBF, but others without CBF showed EEG activity, too.[3,4,11,37–39] Studies have found that one third of newborns with isoelectric EEG showed evidence of CBF and 58% of those with absent CBF had evidence of EEG activity.[4]

Ashwal concluded that BD can be diagnosed in the full-term newborn even if younger than 7 days of age. Due to the limitations of both CBF determinations and EEG findings for BD confirmation in neonates, an observation period of 48 hours is recommended to confirm the diagnosis. If an EEG is isoelectric or if a CBF study shows no flow, the observation period can be shortened to 24 hours. Although brain-dead preterm infants are few in number, the same period of observations is recommended. Ashwal reported that in following these guidelines, the risk of misdiagnosis appears to be exceedingly low.[4,11–15,19,39]

Anencephalics

The use of anencephalic infants as organ donors has generated considerable controversy,[18,40–80] especially when the American Medical Association Council report (which was subsequently withdrawn) recommended that live-born anencephalic infants who did not meet accepted BD criteria could be considered for organ donation.[81]

According to Shewmon,[82] "anencephaly is generally defined as the congenital absence of the skull, scalp, and forebrain (cerebral hemispheres)." Although in most cases the etiology is unknown, the teratogenic factor must be operative in the first month of pregnancy, generating a failure of cranial neurulation.[82–92]

The brainstem exhibits a spectrum of abnormalities that vary from relatively normal to entirely absent.[82,93,94] According to Bernat,[93] approximately half of *anencephalics* who reach the end of pregnancy will be stillborn, depending precisely on the structural abnormalities of the brainstem.

A wide range of neurologic dysfunction can occur, as the brainstem structural abnormalities vary from severe to relatively mild. The neural structures that mediate typical newborn behaviors are located mainly in the brainstem; therefore, anencephalics with relatively intact brainstems exhibit behaviors such as purposeless back-and-forth movements of the extremities, sucking and swallowing, normal orofacial expressions to gustatory stimuli, crying, withdrawal from noxious stimuli, and wake/sleep cycles, which could be misunderstood by parents as normal functioning.[82,93,94] Because of these behavioral similarities between anencephalics and normal newborns, it has been suggested that although the cerebral hemispheres are apparently developed macroscopically in normal newborns, they are very immature microscopically compared to the brainstem.[82] Patterns of reduced regional cerebral metabolism, demonstrated by positron emission tomography in an older child or adult, would constitute a convincing evidence for a persistent vegetative state (PVS). Similar patterns are found in normal human neonates, which suggest a relative lack of normal cerebral cortical function.[82,95,96] "In humans, the cerebral cortex assumes its adult-type primacy in the functional hierarchy of the brain, only with the developmental process known as encephalization."[82] Taking these facts into consideration, several authors have emphasized that anencephalic neonates are not simply "miniature PVS patients."[82,91,94]

Another clue is the potential plasticity of the brainstem in newborns. Clinical and experimental evidence convincingly suggests that the brainstem of newborns is potentially capable of much more complex integrative functioning than is usually imputed to it, including some functions commonly considered to be cortical, even in animals.[82,94,97–101]

According to Shewmon,[82] whether or not an anencephalic neonate, or even a normal infant, is "conscious" or capable of suffering is a philosophical question, empirically unanswerable, but the functional capacity of the brainstem in the early periods of life has led some authors to accept the potential presence of some primitive form of awareness in anencephalics, and the possibility of the subjective feeling of pain.[82,102–108]

The prognosis and causes of death have been extensively studied in anencephalic infants.[82,93,94,109–114] It has been commonly stated that these infants uniformly die no later than a few days after birth, although there have been also documented cases of infants living for months.[110,112,115] The causes of death are related to the abnormalities of the brain and other organ systems, but it is improbable that the primary cause of death of these neonates is brainstem destruction or degeneration.[82,93,110]

Thus, if controversies persist about the possible presence of some primitive form of awareness in anencephalics and of subjective experience of pain,[82,102–108] it becomes difficult to use these infants as organ donors.[2,6,116–119]

However, according to Ashwal,[3] BD can be diagnosed in anencephalic infants by specifically determining the absence or loss of preexisting brainstem function including apnea. It is necessary to perform repetitive examinations to demonstrate the loss of any previously detectable cranial nerve reflexes, and to perform repeated apnea challenge studies to document the absence of respiratory effort during apnea tests. It is recommended that the period of observation to confirm the diagnosis of brain death be similar, that is, 24 to 48 hours.

From the practical (not conceptual) point of view, Shewmon and his colleagues[82,94] have argued that in the United States a very low number of successful transplantation from anencephalics would be expected. Bernat[93] stated, "There has appeared to be a gathering consensus that the use of anencephalics as organ donors is misguided and should be abandoned, and that to alter our concept and statutory definition of death to encompass anencephalic neonates with organ transplantation, is to disrupt a beneficial and coherent framework."

Fetal Brain Death

Adams et al.[120] first reported a brain-dead fetus in 1977. They based their diagnosis on the use of clinical diagnostic criteria for neonates, and found neuropathologic confirmation of diffuse CNS destruction typical of severe intrauterine anoxia and circulatory failure. Fetal brain death is recognized by the presence of a prolonged fixed fetal heart rate pattern, even after applying different types of stimuli for increasing the heart rate (even following different types of stress tests). Fetal brain death is noted in an atonic and apneic fetus without any body movement, in the appearance of polyhydramnios, and in the development of ventriculomegaly.[120–124] Fetal brain death during pregnancy is an extremely rare event, and thus few cases have been reported in the literature.[125]

References

1. Machado C. [Resolution for the determination and certification of death in Cuba]. *Rev Neurol.* 2003;36:763–770.
2. Machado-Curbelo C. [Do we defend a brain oriented view of death?]. *Rev Neurol.* 2002;35:387–396.
3. Ashwal S. Clinical diagnosis and confirmatory testing of brain death in children. In: Wijdicks EFM, ed. *Brain Death*. Philadelphia: Lippincott Williams & Wilkins, 2001:91–114.
4. Ashwal S. Brain death in the newborn. Current perspectives. *Clin Perinatol.* 1997;24:859–882.
5. Ashwal S. Brain death in children and related issues. In: Machado C, ed. *Brain Death (Proceedings of the Second International Symposium on Brain Death)*. Amsterdam: Elsevier Science BV, 1995:201–206.
6. Machado C. Cerebral processing in the minimally conscious state. *Neurology.* 2005;65:973–974.
7. Task Force for the Determination of Brain Death in Children. Guidelines for the determination of brain death in children. *Pediatr Neurol.* 1987;3:242–243.
8. Mejia RE, Pollack MM. Variability in brain death determination practices in children. *JAMA.* 1995;274:550–553.
9. Tsai WH, Lee WT, Hung KL. Determination of brain death in children—a medical center experience. *Acta Paediatr Taiwan.* 2005;46:132–137.
10. Guidelines for the determination of brain death in children. Task Force for the determination of brain death in children. *Neurology.* 1987;37:1077–1078.
11. Ashwal S, Serna-Fonseca T. Brain death in infants and children. *Crit Care Nurse.* 2006;26:117–128.
12. Michelson DJ, Ashwal S. Evaluation of coma and brain death. *Semin Pediatr Neurol.* 2004;11:105–118.
13. Ashwal S, Schneider S. Brain death in children: Part I. *Pediatr Neurol.* 1987;3:5–11.
14. Ashwal S, Schneider S. Brain death in children: Part II. *Pediatr Neurol.* 1987;3:69–77.
15. Ashwal S. Brain death in early infancy. *J Heart Lung Transplant.* 1993;12:S176–S178.
16. Ashwal S, Rust R. Child neurology in the 20th century. *Pediatr Res.* 2003;53:345–361.
17. ten Berge J, de Gast-Bakker DA, Plotz FB. Circumstances surrounding dying in the paediatric intensive care unit. *BMC Pediatr* 2006;6:22.

18. Ashwal S, Schneider S. Pediatric brain death: current perspectives. *Adv Pediatr.* 1991;38:181–202.
19. Drake B, Ashwal S, Schneider S. Determination of cerebral death in the pediatric intensive care unit. *Pediatrics.* 1986;78:107–112.
20. Sheikh AA, Cusack DA. Maternal brain death, pregnancy and the foetus: the medico-legal implications for Ireland. *Med Law.* 2004;23(2):237–245.
21. Ruiz-Garcia M, Gonzalez-Astiazaran A, Collado-Corona MA, Rueda-Franco F, Sosa-de-Martinez C. Brain death in children: clinical, neurophysiological and radioisotopic angiography findings in 125 patients. *Childs Nerv Syst.* 2000;16:40–45.
22. Shewmon DA. Brain death in children. *Neurology.* 1988;38:1813–1814.
23. Guidelines for the determination of brain death in children. Task Force for the Determination of Brain Death in Children. *Arch Neurol.* 1987;44:587–588.
24. Rowland TW, Donnelly JH, Jackson AH. Apnea documentation for determination of brain death in children. *Pediatrics.* 1984;74:505–508.
25. Parvey LS, Gerald B. Arteriographic diagnosis of brain death in children. *Pediatr Radiol.* 1976;4:79–82.
26. Paret G, Barzilay Z. Apnea testing in suspected brain dead children—physiological and mathematical modelling. *Intensive Care Med.* 1995;21:247–252.
27. Chantarojanasiri T, Preutthipan A. Apnea documentation for determination of brain death in Thai children. *J Med Assoc Thai.* 1993;76(suppl 2):165–168.
28. Outwater KM, Rockoff MA. Apnea testing to confirm brain death in children. *Crit Care Med.* 1984;12:357–358.
29. Vardis R, Pollack MM. Increased apnea threshold in a pediatric patient with suspected brain death. *Crit Care Med.* 1998;26:1917–1919.
30. Lang CJ, Heckmann JG. Apnea testing for the diagnosis of brain death. *Acta Neurol Scand.* 2005;112:358–369.
31. Lang CJ, Heckmann JG. How should testing for apnea be performed in diagnosing brain death? *Adv Exp Med Biol.* 2004;550:169–174.
32. Lang CJ, Heckmann JG, Erbguth F et al. Transcutaneous and intra-arterial blood gas monitoring—a comparison during apnoea testing for the determination of brain death. *Eur J Emerg Med.* 2002;9:51–56.
33. Lang CJ. Apnea testing guided by continuous transcutaneous monitoring of partial pressure of carbon dioxide. *Crit Care Med.* 1998;26:868–872.
34. Lang CJ. Blood pressure and heart rate changes during apnoea testing with or without CO_2 insufflation. *Intensive Care Med.* 1997;23:903–907.
35. Lang CJ. Apnea testing by artificial CO_2 augmentation. *Neurology.* 1995;45:966–969.
36. Ashwal S, Schneider S. Brain death in the newborn. *Pediatrics.* 1989;84:429–437.
37. Ashwal S, Schneider S. Failure of electroencephalography to diagnose brain death in comatose children. *Ann Neurol.* 1979;6:512–517.
38. Schneider S. Usefulness of EEG in the evaluation of brain death in children: the cons. *Electroencephalogr Clin Neurophysiol.* 1989;73:276–278.
39. Alvarez LA, Moshe SL, Belman AL, Maytal J, Resnick TJ, Keilson M. EEG and brain death determination in children. *Neurology.* 1988;38:227–230.
40. Machado C. Death on neurological grounds. *J Neurosurg Sci.* 1994;38:209–222.
41. Ashwal S. Recovery of consciousness and life expectancy of children in a vegetative state. *Neuropsychol Rehabil.* 2005;15:190–197.
42. Ford NM. Newborns and organ donation: some guidelines for decision making. *Ethics Medics.* 2003;28:2–4.
43. Veatch RM. The dead donor rule: true by definition. *Am J Bioeth.* 2003;3:10–11.

44. Schlotzhauer AV, Liang BA. Definitions and implications of death. *Hematol Oncol Clin North Am.* 2002;16:1397–1413.
45. Spinney L. The living dead. *New Sci.* 2001;171:38–41.
46. Pasquerella L, Smith S, Ladd R. Infants, the dead donor rule, and anencephalic organ donation: should the rules be changed? *Med Law.* 2001;20:417–423.
47. Bard JS. The diagnosis is anencephaly and the parents ask about organ donation: now what? A guide for hospital counsel and ethics committees. *West New Engl Law Rev.* 1999;21:49–95.
48. Robertson JA. The dead donor rule. *Hastings Cent Rep.* 1999;29:6–14.
49. Cady R. Anencephalics as organ donors: where do we stand? *MCN Am J Matern Child Nurs.* 1999;24:51.
50. Simini B. Revolutionary swap of baby organs ends badly. *Lancet.* 1998;351:503.
51. Lafreniere R, McGrath MH. End-of-life issues: Anencephalic infants as organ donors. *J Am Coll Surg.* 1998;187:443–447.
52. Walters J, Ashwal S, Masek T. Anencephaly: where do we now stand? *Semin Neurol.* 1997;17:249–255.
53. Abbattista AD, Vigevano F, Catena G, Parisi F. Anencephalic neonates and diagnosis of death. *Transplant Proc.* 1997;29:3634–3635.
54. Shewmon DA. Recovery from "brain death": a neurologist's apologia. *Linacre Q.* 1997;64:30–96.
55. Rhodes AM. Testing the standards of death. *MCN Am J Matern Child Nurs.* 1996;21:109.
56. Hanger LE. The legal, ethical, and medical objections to procuring organs from anencephalic infants. *Health Matrix Cleve.* 1995;5:347–368.
57. Khan JH. Anencephalic infants as organ donors. *JAMA.* 1995;274:1758–1759.
58. Justice JS. Personhood and death—the proper treatment of anencephalic organ donors under the law: In re T.A.C.P., 609 So. 2d 588 (Fla. 1992). *University of Cincinnati Law Rev.* 1994;62:1227–1279.
59. Eitzman DV. Use of anencephalics as organ donors. *J Fla Med Assoc.* 1994;81: 27–29.
60. Girvin J, Capron AM. Critical issues debates: intervention for infants with fatal heart disease, xenografting, and brain death criteria for anencephalic infants. Debate III. Resolved: brain death criteria must be revised so that society can readily benefit from families who offer their anencephalic infants as organ donors. *J Heart Lung Transplant.* 1993;12:S369–S378.
61. Hetzer R, Franz N. [Organ donation of moribund newborn infants from the viewpoint of transplantation surgery]. *Z Arztl Fortbild (Jena).* 1993;87:887–890.
62. Lizza JP. Persons and death: what's metaphysically wrong with our current statutory definition of death? *J Med Philos.* 1993;18:351–374.
63. Donovan GK. Anencephalic organ donation in Oklahoma. Right problem, wrong answer. *J Okla State Med Assoc.* 1993;86:128–130.
64. Koenig J. The anencephalic Baby Theresa: a prognosticator of future bioethics. *Nova Law Rev.* 1992;17:445–496.
65. Cranford RE. Anencephalic infants as organ donors. *Transplant Proc.* 1992;24:2218–2220.
66. Infants with anencephaly as organ sources: ethical considerations. *Pediatrics.* 1992;89:1116–1119.
67. Theresa Ann Campo Pearson: baby girl. *New York Times* 1992;A20. July 29, 1992.
68. Ahmad F. Anencephalic infants as organ donors: beware the slippery slope. *Can Med Assoc J.* 1992;146:236–241, 244.

69. Paris JJ, Signorello G. The use of anencephalic organ donors: lesson of Baby Theresa Ann. *Clin Ethics Rep.* 1992;6:3–6.

70. Peabody JL. Reflections on the Loma Linda University experience. *Clin Ethics Rep.* 1992;6:1–2.

71. Kushner T. When do organs become "spare parts"? *Camb Q Healthcare Ethics.* 1992;1:349–353.

72. Ashwal S, Caplan AL, Cheatham WA, Evans RW, Peabody JL, Larson D. Session IX: Social and ethical controversies in pediatric heart transplantation. *J Heart Lung Transplant.* 1991;10:860–876.

73. Walters JW. Anencephalic infants as organ sources. *Bioethics.* 1991;5:326–341.

74. Sommerauer JF. Brain death determination in children and the anencephalic donor. *Clin Transplant.* 1991;5:137–145.

75. Taylor RM. Anencephalic donors and renal transplantation. *Br J Hosp Med.* 1991;45:136.

76. Tonti-Filippini N. Determining when death has occurred. *Linacre Q.* 1991;58:25–49.

77. Ashwal S, Peabody JL, Schneider S, Tomasi LG, Emery JR, Peckham N. Anencephaly: clinical determination of brain death and neuropathologic studies. *Pediatr Neurol.* 1990;6:233–239.

78. Rast AM. Anencephalic infants and organ procurement. *Imprint.* 1990;37:61–62.

79. Neonatal heart transplants: the ethical problems. *Brief Med Ethics.* 1990;No. 6: 1–4.

80. May WF. Brain death: anencephalics and aborted fetuses. *Transplant Proc.* 1990;22:985–988.

81. The use of anencephalic neonates as organ donors. Council on Ethical and Judicial Affairs, American Medical Association. *JAMA.* 1995;273:1614–1618.

82. Shewmon DA. Anencephaly: selected medical aspects. *Hastings Cent Rep.* 1988;18:11–19.

83. Botkin JR. Anencephalic infants as organ donors. *Pediatrics.* 1988;82:250–256.

84. Rogner UC, Spyropoulos DD, Le Novere N, Changeux JP, Avner P. Control of neurulation by the nucleosome assembly protein-1–like 2. *Nat Genet.* 2000;25:431–435.

85. Brook FA, Estibeiro JP, Copp AJ. Female predisposition to cranial neural tube defects is not because of a difference between the sexes in the rate of embryonic growth or development during neurulation. *J Med Genet.* 1994;31:383–387.

86. Savel'ev SV, Chernikov VP. [The morphogenesis of the human brain on the 27th-35th day of development with disordered neurulation]. *Izv Akad Nauk Ser Biol.* 1992;465–471.

87. Wood LR, Smith MT. Generation of anencephaly: 1. Aberrant neurulation and 2. Conversion of exencephaly to anencephaly. *J Neuropathol Exp Neurol.* 1984;43:620–633.

88. Silverman HJ. Withdrawal of feeding-tubes from incompetent patients: the Terri Schiavo case raises new issues regarding who decides in end-of-life decision making. *Intensive Care Med.* 2005;31:480–481.

89. Finnell RH, Junker WM, Wadman LK, Cabrera RM. Gene expression profiling within the developing neural tube. *Neurochem Res.* 2002;27:1165–1180.

90. Van Allen MI, Kalousek DK, Chernoff GF et al. Evidence for multi-site closure of the neural tube in humans. *Am J Med Genet.* 1993;47:723–743.

91. Park CH, Stewart W, Khoury MJ, Mulinare J. Is there etiologic heterogeneity between upper and lower neural tube defects? *Am J Epidemiol.* 1992;136:1493–1501.

92. Seller MJ. Neural tube defects and sex ratios. *Am J Med Genet.* 1987;26:699–707.

93. Bernat JL. *Ethical Issues in Neurology.* Philadelphia: Lippincott, 1991.

94. Shewmon DA, Capron AM, Peacock WJ, Schulman BL. The use of anencephalic infants as organ sources. A critique. *JAMA.* 1989;261:1773–1781.

95. Latronico N, Alongi S, Guarneri B, Cappa S, Candiani A. [Approach to the patient in vegetative state. Part I: diagnosis]. *Minerva Anestesiol.* 2000;66:225–231.

96. The Medical Task Force on Anencephaly. The infant with anencephaly. *N Engl J Med.* 1990;322:669–674.

97. Hovda DA, Sutton RL, Feeney DM. Amphetamine-induced recovery of visual cliff performance after bilateral visual cortex ablation in cats: measurements of depth perception thresholds. *Behav Neurosci.* 1989;103:574–584.

98. Hovda DA, Sutton RL, Feeney DM. Recovery of tactile placing after visual cortex ablation in cat: a behavioral and metabolic study of diaschisis. *Exp Neurol.* 1987;97:391–402.

99. Feeney DM, Hovda DA. Reinstatement of binocular depth perception by amphetamine and visual experience after visual cortex ablation. *Brain Res.* 1985;342:352–356.

100. Walker AE, Feeney DM, Hovda DA. The electroencephalographic characteristics of the rhombencephalectomized cat. *Electroencephalogr Clin Neurophysiol.* 1984;57:156–165.

101. Feeney DM, Hovda DA. Amphetamine and apomorphine restore tactile placing after motor cortex injury in the cat. *Psychopharmacology (Berl).* 1983;79:67–71.

102. Bjursten LM, Norrsell K, Norrsell U. Behavioural repertory of cats without cerebral cortex from infancy. *Exp Brain Res.* 1976;25:115–130.

103. Bartocci M, Bergqvist LL, Lagercrantz H, Anand KJ. Pain activates cortical areas in the preterm newborn brain. *Pain.* 2006;122(1–2):109–117.

104. Tibboel D, Anand KJ, van den Anker JN. The pharmacological treatment of neonatal pain. *Semin Fetal Neonatal Med.* 2005;10:195–205.

105. Anand KJ. Pain, plasticity, and premature birth: a prescription for permanent suffering? *Nat Med.* 2000;6:971–973.

106. Anand KJ. Neonatal analgesia and anesthesia. Introduction. *Semin Perinatol.* 1998;22:347–349.

107. Anand KJ, Carr DB. The neuroanatomy, neurophysiology, and neurochemistry of pain, stress, and analgesia in newborns and children. *Pediatr Clin North Am.* 1989;36:795–822.

108. Anand KJ, Hickey PR. Pain and its effects in the human neonate and fetus. *N Engl J Med.* 1987;317:1321–1329.

109. Peabody JL, Emery JR, Ashwal S. Experience with anencephalic infants as prospective organ donors. *N Engl J Med.* 1989;321:344–350.

110. Baird PA, Sadovnick AD. Survival in liveborn infants with anencephaly. *Am J Med Genet.* 1987;28:1019–1020.

111. Ekinci G, Balci S, Erzen C. An anencephalic monocephalus diprosopus "headed twin": postmortem and CT findings with emphasis on the cranial bones. *Turk J Pediatr.* 2005;47:195–198.

112. Forrester MB, Merz RD. First-year mortality rates for selected birth defects, Hawaii, 1986–1999. *Am J Med Genet A.* 2003;119:311–318.

113. Dai L, Zhu J, Zhou GX et al. [Clinical features of 3798 perinatals suffering from syndromic neural tube defects]. *Zhonghua Fu Chan Ke Za Zhi.* 2003;38:17–19.

114. Dai L, Zhu J, Zhou G et al. [Dynamic monitoring of neural tube defects in China during 1996 to 2000]. *Zhonghua Yu Fang Yi Xue Za Zhi.* 2002;36:402–405.

115. Baird PA, Sadovnick AD. Survival in infants with anencephaly. *Clin Pediatr (Phila).* 1984;23:268–271.

116. Machado C. Can vegetative state patients retain cortical processing? *Clin Neurophysiol.* 2005;116:2253–2254.

117. Machado C, Sherman DL. *Brain Death and Disorders of Consciousness.* New York: Kluwer Academics/Plenum, 2004.

118. Machado C. Is the concept of brain death secure? In: Zeman A, Enamorado A, eds. *Ethical Dilemmas in Neurology.* London: WB Saunders, 2000:193–212.

119. Machado C. Consciousness as a definition of death: its appeal and complexity. *Clin Electroencephalogr.* 1999;30:156–164.

120. Adams RD, Prod'hom LS, Rabinowicz T. Intrauterine brain death. Neuraxial reticular core necrosis. *Acta Neuropathol (Berl).* 1977;26(40):41–49.

121. Zimmer EZ, Jakobi P, Goldstein I, Gutterman E. Cardiotocographic and sonographic findings in two cases of antenatally diagnosed intrauterine fetal brain death. *Prenat Diagn.* 1992;12:271–276.

122. Powner DJ, Bernstein IM. Extended somatic support for pregnant women after brain death. *Crit Care Med.* 2003;31:1241–1249.

123. James SJ. Fetal brain death syndrome—a case report and literature review. *Aust N Z J Obstet Gynaecol.* 1998;38:217–220.

124. Nijhuis JG, Crevels AJ, van Dongen PW. Fetal brain death: the definition of a fetal heart rate pattern and its clinical consequences. *Obstet Gynecol Surv.* 1990;45:229–232.

125. Marquis D. Abortion and the beginning and end of human life. *J Law Med Ethics.* 2006;34(1):16–25.

7
Vegetative and Minimally Conscious States and Other Disturbances of Consciousness

The Terry Schiavo case has raised new controversies about the diagnosis and management of the persistent vegetative state (PVS) and the minimally conscious state (MCS).[1-12] This controversy dominated the national news in the United States for some time and the case was taken to the courts, the Florida legislature, the Florida governor, the Congress, and all the way to the President of the United States.[4,13-18] The Schiavo case and other famous patients, including Karen Ann Quinlan[19-31] and Nancy Cruzan in the United States[32-43] and Tony Bland in the United Kingdom,[44-47] have made it necessary for neurologists and neuroscientists to propose reliable diagnostic guidelines for testing brain function in altered states of consciousness.[48-53]

The term *persistent vegetative state* was coined by Jennett and Plum in 1972 to describe the condition of patients with severe brain damage in whom coma has progressed to a state of wakefulness without detectable awareness. Such patients have sleep-wake cycles but no ascertainable cerebral cortical function. Jennett and Plum thought that patients in a persistent vegetative state could be distinguished clinically from those with other conditions associated with disturbances of consciousness. They described a specific syndrome of reflex reactions without any meaningful response to the environment but in patients who have a sleep-wake pattern. They introduced the term because they were dissatisfied with the other terms used at the time. These terms referred to states that were not correct, for example, prolonged *coma* or *coma* vigil (the patients, by definition, were not in a coma), or described specific syndromes, for example, decerebrate dementia, parasomnia, or akinetic mutism.[54] The term *apallic syndrome*, still used in some European countries, implies lack of the cortex (pallium), but this is not the only pathologic pattern found in these cases.[55-60]

The main finding in PVS is preservation of wakefulness with apparent loss of awareness.[61-69] The diagnosis of PVS has been made more difficult by recognition of the minimally conscious state (MCS) as a transitional phase in the partial recovery of self-awareness or environmental awareness while emerging from the PVS, leading to a relatively high proportion of errors.[67,68,70-75]

Clinical Findings in Persistent Vegetative State Patients

Diagnostic criteria for PVS include the lack of evidence of awareness of self or environment, a lack of interaction with others, and the lack of comprehension or expression of language. By implication, external stimuli do not evoke purposeful or sustained and reproducible voluntary behavioral responses.[50,52,70,76-84] The accurate diagnosis of PVS requires the skills of a multidisciplinary team experienced in the management of people with complex disabilities and consciousness disorders. Accurate clinical assessments of patients in these conditions must be obtained before they undergo neuroimaging. Moreover, in reports of neuroimaging studies, all relevant clinical details must be available for comparisons between studies.[65,78,85]

The Multi-Society Task Force on PVS has classified the causes of PVS in three main groups: acute injuries, in which the most common causes are traumatic and hypoxic-ischemic encephalopathy; degenerative and metabolic disorders, including dementia; and developmental malformations, of which the most important is anencephaly. Nonetheless, the most prevalent causes of acute PVS at all ages are head trauma and hypoxic-ischemic encephalopathy. These causes have been taken as models to describe the three main patterns of the neuropathological damage in PVS cases.[28,86] The Multi-Society Task Force on PVS has defined the precise use of the terms *persistent* and *permanent:* "Persistent refers only to a condition of past and continuing disability with an uncertain future, whereas permanent implies irreversibility." This Task Force likewise stated that "a patient in a persistent vegetative state becomes permanently vegetative when the diagnosis of irreversibility can be established with a high degree of clinical certainty." According to the etiology, a period of observation has been proposed to define when a persistent vegetative state has become a permanent vegetative state.[87,88] When the term *persistent vegetative state* was first described, it was emphasized that persistent did not mean permanent; it is now recommended that "persistent" be omitted and patients be described as having been vegetative for a certain time.[78]

According to the Multi-Society Task Force on PVS,[87,88] the vegetative state can be diagnosed considering the following criteria listed in Table 7.1. These criteria include the lack of awareness of self or environment and the inability to interact with others, the lack of sustained, reproducible, purposeful, or voluntary behavioral responses to visual auditory, tactile, or noxious stimuli, and a lack of evidence of language comprehension or expression. There is intermittent wakefulness manifested by the presence of sleep-wake cycles, a sufficient hypothalamic and brainstem autonomic function for survival with medical and nursing care, bowel and bladder incontinence, and variably preserved cranial-nerve reflexes and spinal reflexes.[62,70,76,78,87-89] The crux of the problem lies in the determination of a person's internal mental state using external proof.[67-69]

Eye tracking and emotional responses are the most common ways of determining whether a patient is responding and therefore no longer in a vegetative state.

TABLE 7.1. Criteria for diagnosing the vegetative state, according to the Multi-Society Task Force on persistent vegetative state (PVS)

1. No evidence of awareness of self or environment and an inability to interact with others
2. No evidence of sustained, reproducible, purposeful, or voluntary behavioral responses to visual, auditory, tactile, or noxious stimuli
3. No evidence of language comprehension or expression
4. Intermittent wakefulness manifested by the presence of sleep-wake cycles
5. Sufficiently preserved hypothalamic and brainstem autonomic functions to permit survival with medical and nursing care
6. Bowel and bladder incontinence
7. Variably preserved cranial-nerve reflexes (pupillary, oculocephalic, corneal, vestibulo-ocular, and gag) and spinal reflexes

The first sign of a patient emerging from PVS is the localizing of the eyes on a visual stimulus. This can be observed because persons in a vegetative state are unable to track moving objects or fixate their vision on an object, and as patients recover they regain this ability.[89–94] Several authors have reported that recovery of visual pursuit may presage recuperation of other signs of consciousness.[95,96] Giacino and Kalmar[80] reported a significantly higher incidence of visual pursuit among MCS patients admitted for inpatient rehabilitation, independent of time postinjury. Additionally, among the 20% of vegetative state (VS) patients who exhibited visual pursuit, 73% showed clear-cut signs of consciousness by 12 months postinjury, while only half of the VS patients without pursuit movements emerged from VS.

However, Andrews[90] emphasized that many patients who are misdiagnosed as being in the vegetative state are blind or have severe visual handicap, and the lack of an eye blink to threat or the absence of visual tracking might not be reliable signs for diagnosing the vegetative state.[90]

Giacino et al.[50–52] developed the JFK Coma Recovery Scale to follow up the clinical evolution in these patients and their potential recovery. If the Coma Recovery Scale-revised (CRS-R) is applied by trained examiners to derive repeated measurements, it could yield stable estimates of patient status. This scale contains six subscales: auditory function, visual function, motor function, oromotor/verbal function, communication, and arousal scales.

Misdiagnoses of the vegetative state, minimally conscious state, and locked-in syndrome are common. Unexpected and well-documented recoveries of cognitive functions have been described in patients diagnosed by neurologists experienced and skilled in the diagnosis of this condition.[50,64,79–81,91,97–98,99–102] Childs et al.[99] reported that 37% of 49 cases admitted to a special unit for rehabilitation were incorrectly diagnosed, according to the American Medical Association guidelines on PVS. Although imaging techniques have the potential to improve both diagnostic and prognostic accuracy, careful and repeated neurological assessment by a trained examiner remains the best practice.[99]

Comparison of Brain Death, Coma, Vegetative State, Minimally Conscious State, and Locked-in Syndrome

Brain Death

This state has been discussed in earlier chapters.

Coma

Coma is a sleep-like state, characterized by the absence of arousal and awareness. The patient lies with the eyes closed, cannot be aroused, and has no awareness of self and surroundings. Stimulation cannot produce spontaneous periods of wakefulness or eye opening in patients in a coma, unlike patients in a vegetative state. To have a clear distinction from other states of transient loss of consciousness, such as syncope and concussion, the coma must last for at least 1 hour.[48,52,77,78,84,94,103-104,105,107] Coma can result from diffuse bihemispheric cortical or white-matter damage after neuronal or axonal injury, or from focal brainstem lesions that affect the pontomesencephalic tegmentum or paramedian thalami bilaterally.[104,108]

Locked-In Syndrome

The locked-in syndrome is characterized by sustained eye opening (bilateral ptosis should be ruled out as a complicating factor), aphonia or hypophonia, quadriplegia or quadriparesis, and vertical or lateral eye movement or blinking of the upper eyelid to signal yes/no responses, but patients preserve awareness of the environment. Eye or eyelid movements are the main method of communication. This state entails a serious issue for the medical staff, because patients can hear and understand comments about their critical health state. The term *locked-in syndrome* was coined by Plum and Posner to describe the quadriplegia and anarthria that are a consequence of the disruption of corticospinal and corticobulbar pathways.[108]

Minimally Conscious State

Minimally conscious state (MC; Table 7.2) patients show limited but clear evidence of awareness of themselves or their environment, based on a reproducible or sustained basis, by at least one of the following behaviors: gestural or verbal yes/no response (regardless of accuracy), obeying simple commands, intelligible speech, and purposeful behavior (including movements or affective behavior that happen in response to stimuli in the environment and are not due to reflexive activity).[67,68,70-73,77,79-80,83,97,109] The Aspen group proposed the criteria for minimally conscious state to categorize patients who are not in a vegetative state but are unable to communicate consistently.[97] The MCS in some patients is a transitional phase toward the partial recovery of self-awareness or environmental awareness while emerging from the VS.[70-72,80,97]

TABLE 7.2. Diagnostic criteria for minimally conscious state (MCS), to demonstrate clear and discernible evidence of self or environmental awareness, on a reproducible or sustained basis

1. Obeying simple commands
2. Gestural or verbal yes/no responses (regardless of accuracy)
3. Intelligible verbalization
4. Purposeful behavior, including movements or affective behaviors that occur in contingent relation to relevant environmental stimuli and are not due to reflexive activity; some examples of qualifying purposeful behavior include:
   - Appropriate smiling or crying in response to the linguistic or visual content of emotional but not to neutral topics or stimuli
   - Vocalizations or gestures that occur in direct response to the linguistic content of questions
   - Reaching for objects that demonstrates a clear relationship between object location and direction of reach
   - Touching or holding objects in a manner that accommodates the size and shape of the object
   - Pursuit eye movement or sustained fixation that occurs in direct response to moving or salient stimuli

Emergence from the MCS is defined by the ability to communicate or use objects functionally. Further improvement is more likely than in patients in a vegetative state, although some patients remain in an MCS permanently.[80,97]

Giacino et al.[97] proposed a group of diagnostic criteria for MCS, to demonstrate clear and discernible evidence of self or environmental awareness, on a reproducible or sustained basis (Table 7.2).

Akinetic mutism was first described in Cairn in a girl with a craniopharyngiomatous cyst that compressed the walls of the third ventricle.[110] Akinetic mutism is a rare state that has been described as a subcategory of the MCS, because although the syndrome has fairly identifiable characteristics, there is a continuum and overlap with MCS.[111,112]

The main clinical features are an abulic emotionless state, unresponsiveness, and eye tracking of objects or persons in the hospital room or intensive care unit. Acording to Wijdicks and Cranford,[111] the description of this syndrome is difficult because most patients with akinetic mutism are not fully akinetic, and many respond with movements. Also, a few patients are mute, and many are able to utter only a single word. Wikdicks and Cranford also stated that the capability of the patient visually to follow the examiner in the room or abruptly be prompted by a person entering the room has been compared to an observable condition detected in monkeys after the temporal lobes have been removed, known as hypermetamorphosis; these animals give the impression of responding to movement rather than to objects.[111]

It has been suggested that this syndrome is caused by bilateral lesions of the cingulate gyri. The anterior cingulate cortex lesions affect executive functions and vocalization and, as this area is highly connected with the supplementary motor area, impair initiation of movement.[113] Nonetheless, other brain lesions have been reported to explain this syndrome, for example, lesions affecting the diencephalic

TABLE 7.3. Comparison among brain death (BD), coma, persistent vegetative state (PVS), minimally conscious state (MCS), and locked-in state

Features	BD	Coma	PVS	MCS	Locked-in
Arousal	No	No*	Yes	Yes	Yes
Awareness	No	No**	No	Yes (partially preserved)	Yes
Time for diagnosis	Hours or days	Hours or days	Months	Months	Weeks or months
Eye opening	Never, even on noxious stimulation	Never, even on noxious stimulation	Spontaneously open	Spontaneously open	Spontaneously open
Apnea	Apneic; Require controlled artificial ventilation	Artificial ventilation is required, although patients can make respiratory efforts	Patients can breathe spontaneously without assistance, or artificial ventilation is used only for support	Patients can breathe spontaneously without assistance, or artificial ventilation is used only for support	Patients can breathe spontaneously without assistance, or artificial ventilation is used only for support
Brainstem function	Absent		Preserved	Preserved	Preserved corticobulbar pathways are damaged
Hypothalamic function	Absent, rarely preserved		Preserved	Preserved	Preserved
Body movements	Generated by residual spinal activity; Facial myokymia up to a third of patients		Patients show a richer array of motor activity, albeit always nonpurposeful, inconsistent and noncoordinated	Some motor activity might be purposeful, consistent and coordinated	Quadriplegia or quadriparesis
Facial and vocal expression	None		Patients may occasionally smile or cry, utter grunts and sometimes moan or scream	Some facial and vocal expressions might be purposeful, consistent, and coordinated	Vertical or lateral eye movement or blinking of the upper eyelid to signal yes/no responses

* In coma a full arousal response is not preserved, although motor and vegetative reponses are elicited by stimuli.

** Some patients in light stages of coma have referred to hearing voices.

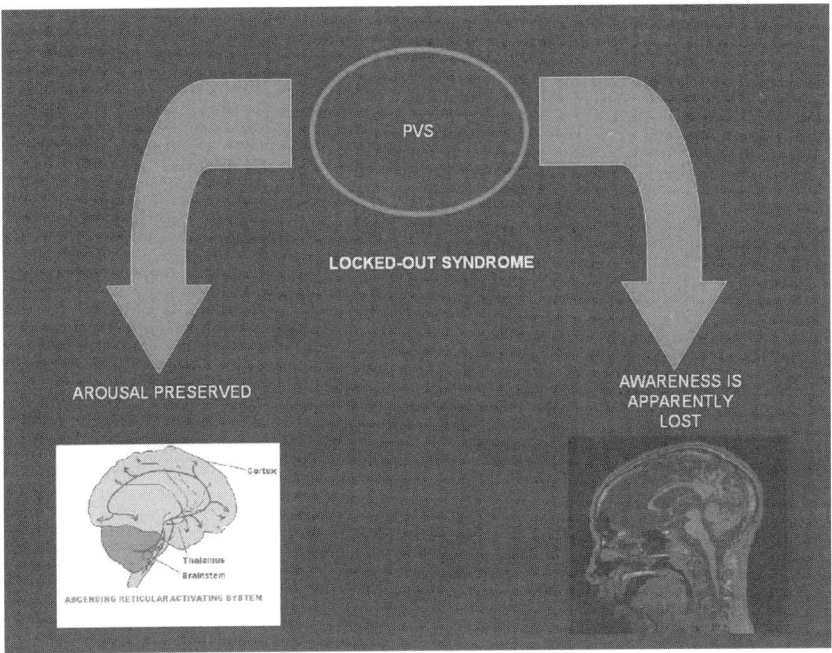

FIGURE 7.1. The main finding in vegetative state (VS) is preservation of wakefulness with apparent loss of awareness. (See also color insert.)

structures, such as the thalamus, basal ganglia, or even mesencephalic structures involving the reticular activating system.[114,115] Akinetic mutism has been described as being associated with hypothalamic lesions and obstructive hydrocephalus,[116] aneurysmal subarachnoid hemorrhage producing bifrontal lesions,[117] and an infiltrative astrocytoma in the fornix[118] (Table 7.3).

Pathophysiological Explanation

Vegetative state patients reflect the only circumstance in which an apparent dissociation of both components of consciousness is found. The main finding in VS is preservation of wakefulness with apparent loss of awareness (Fig. 7.1).[61–69]

The recent description of the MCS in some patients as a transitional phase toward the partial recovery of self-awareness or environmental awareness while emerging from the VS highlights the crucial role of pathophysiological elucidation of consciousness generation in these patients.[28,39,48,50,63,73,80,86,97,109,115,119–135,136–156]

To use the term *unconscious* for characterizing VS might confuse this state with coma. In comatose patients arousal and awareness are lost, but in VS patients arousal is preserved and awareness is aparently lost. Hence, I don't use the terms *awareness* and *consciousness* interchangeably.[87,88]

This raises the question: Why is awareness apparently lacking in VS, while wakefulness is preserved? There is striking evidence that subcortical structures are capable of mediating some form of awareness. Thus, awareness is related not only to the function of the neocortex (although it is primary important), but also to complex physical and psychological mechanisms due to the interrelation of the ascending reticular activating system, thalamus, limbic system, and cerebrum.[65,120,136,137,156–171]

Kretschmer first coined the term *apallic syndrome* to describe the behavior that accompanies the diffuse bilateral cerebral cortex that sometimes follows anoxia head injury and encephalitis. Damage to the cerebral cortex was considered the primary pathological finding.[172] Ingvar[173–175] proposed using this term to describe patients suffering from absent neocortical function but with relatively spared brainstem functioning. However, recent publications have shown that VS results from a complex combination of discrete cortical and subcortical damage. In addition, consciousness requires sufficient thalamocortical and intercortical connections based on work that has shown that isolated functional activity from various parts of the cortex does not provide awareness of self and the environment.[3,4]

This raises the question: Why is awareness lacking in PVS, while arousal is preserved? The neuropathology in the PVS provides a suitable background to discuss the pathophysiology of consciousness generation. Kinney and Samuels[86] presented a detailed review of this subject. According to them, PVS denotes a "locked-out-syndrome" because "the cerebral cortex is disconnected from the external world, and all awareness of the external world is lost." They suggested that the loss of awareness in the PVS is caused by three main patterns: widespread and bilateral lesions of the cerebral cortex, diffuse damage of intra- and subcortical connections in the cerebral hemispheres white matter, and necrosis of the thalamus (Fig. 7.2).

In widespread and bilateral lesions of the cerebral cortex, hypoxic-ischemic encephalopathy is the main etiology. It could be the consequence of acute hypoxic-ischemic insults after cardiorespiratory arrest, strangulation, suffocation, near-drowning, prolonged hypotension, and perinatal asphyxia in neonates.[28–29,50,54,61,63,64,69,78,80,84,86–88,98,100,102,149,156,158,173,176–184] The description of this pattern was the reason that PVS was first known as the apallic syndrome,[55–60,174,185,186] characterized by the destruction of the "pallium, the cortical gray matter that covers the thelencephalon."[186] In the cerebral cortex, a laminar necrosis is found that is multifocal or diffuse and extensive. Other ischemic lesions may be superimposed chiefly in the border zones of the main intracranial cerebral arteries, such as the parasagittal parieto-occipital region, for example.[28,86]

Other damage, such as neuronal loss and small infarcts, is also typically found in the cerebellum, basal ganglia, thalamus, and hippocampus, the latter being particularly sensitive. Other anatomical structures of the brain are relatively undamaged, such as the brainstem, hypothalamus, basal forebrain, and amygdala. This distribution of brain damage reflects the differential vulnerability of brain regions to hypoxia-ischemia.[28,69,86,120,137,139,156,157,187–197]

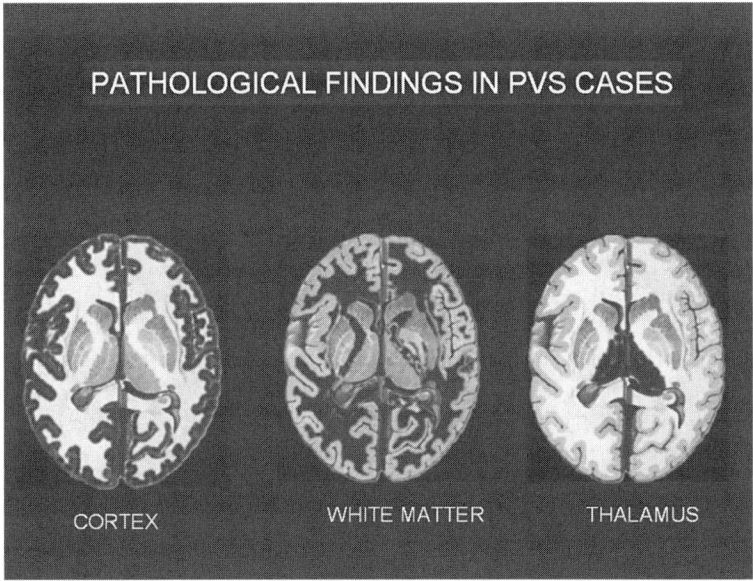

FIGURE 7.2. The VS syndrome is caused by three main patterns: widespread and bilateral lesions of the cerebral cortex, diffuse damage of intra- and subcortical connections in the cerebral hemispheres white matter, and necrosis of the thalamus. (Modified from Kinney et al.[28]) (See also color insert.)

In PVS cases with diffuse damage of the cerebral cortex, the lack of awareness is understandable. The widespread involvement of the association cortices, combined with primary and secondary cortices damage, is the anatomical cause.[28,86] It has been suggested that in diffuse cerebral cortical lesions, the brainstem and the thalamus can maintain arousal.[28,68,69,86,120,137,156,157] Nonetheless, other parallel pathways projecting monosynaptically to the cerebral cortex without relaying through the thalamus could participate in maintaining arousal in these cases. Thus, arousal could be preserved without a functional cerebral cortex. This has been also supported by experimental data. In animals with a total removal of the cerebral cortex or transection at the rostral midbrain level, arousal is preserved, showing sleep-wake cycles.[198] Therefore, it has been argued that the brainstem alone could be sufficient for arousal.[28,86,198]

The mechanism of this pattern could be explained in head trauma and hypoxic-ischemic injury. After head trauma, widespread damage of axons in the cerebral hemispheres white matter occurs, known as diffuse axonal injury (DAI). The DAI is probably caused by the acceleration suffered by the head immediately after the injury.[182,183,184,199-213]

The cerebral hemisphere white matter could also be damaged after hypoxic-ischemic accidents in a pattern known as leukoencephalopathy,[214-217] which is characterized by "extensive symmetrical necrotic lesions in the central white matter of the cerebral hemispheres, with minimal or no damage to gray matter structures."[86] These patients yield antecedents of prolonged periods of hypotension, hypoxemia, and increased venous pressure.[86,182,210]

This pattern also provides a disconnection of the cerebral cortex from the environment that can explain the lack of awareness in the PVS. The functionally unaltered brainstem and thalamus preserve arousal. The participation of other parallel pathways not relaying through the thalamus has also been demonstrated.[86,120,137,156,157]

Reports in PVS patients and experimental data of diffuse axonal injury to cerebral hemispheres, with the cerebral cortex remaining largely normal, suggest that "acute diffuse disconnection of the cerebral cortex from its subcortical activating mechanisms can block arousal as well as cognitive activity in the primate brain."[104]

Another pattern is characterized by a selective necrosis of the thalamus, and although the cortex is not totally spared, the lesions are focal and restricted.[28,86,123–125,128,168,182,183,184,198,199,218–227] It has been explained by several possible factors, such as partial or immediately reversed transtentorial herniation, cerebral edema causing hypoxia-ischemia, and intrinsic metabolic vulnerability of the thalamus.[28,29,31]

The lesions of the thalamus provide a disconnection of the cerebral cortex from the external world, and therefore, all awareness from the environment is lost. The lack of awareness in this pattern is not only a consequence of lesions destroying the sensory relay nuclei that block sensory information from the external world, but the damage of the thalamic intralaminar nuclei is probably the critical anatomical substratum.[28,86,125,222,228–230] These thalamic nuclei receive inputs from many sensory modalities and project over wide areas of the cerebral cortex with a "nondiscernible topography."[28,86] These nuclei integrate important pathways to subserve fundamental cognitive and affective functions, such as attention to the external world.[28,86] It has been argued that lesions in a thalamic nucleus that is preferentially connected with an association cortex provoke functional impairments similar to damage in the association cortex itself.[2,2308] For instance, contrary to the general expectation, the neuropathological examination of Karen Ann Quinlan's brain showed disproportionally severe damage of the thalamus as compared with the cerebral cortex.[19,28–30] Bilateral thalamic infarcts are commonly accompanied by mental impairment, such as dementia and amnesia.[54,217,231–242]

In this pattern arousal could be preserved by a functionally intact brainstem and the other parallel pathways that project to the cerebral cortex, without relaying through the thalamus.[66,137,156,157] It has been argued that "the thalamus is critical for cognition and awareness and may be less essential for arousal."[28]

Neuroimaging Techniques

Functional neuroimaging has provided new insights for assessing cerebral activity in patients with severe brain damage. Measurements of cerebral metabolism and brain activations in response to sensory stimuli with positron emission tomography (PET), functional magnetic resonance imaging (fMRI), and electrophysiological methods can provide information on the presence, degree, and location of any residual brain function in PVS or MCS cases.[61–65,77,78,155,243–246]

Functional imaging with cerebral perfusion tracers and single photon emission tomography (SPECT) or cerebral metabolism tracers and PET typically show a "hollow skull phenomenon" in patients who are brain dead, confirming the absence of neuronal function in the whole brain.[78,194,247––256]

On average, gray-matter metabolism is 50% to 70% of the normal range in comatose patients of traumatic or hypoxic origin.[78,255,257,258] However, in patients with traumatic diffuse axonal injury, both hyperglycolysis and metabolic depression have been reported.[78,259]

Until recently, most studies using PET have shown a substantial reduction in global cerebral glucose metabolism in VS cases below the levels encountered during deep barbiturate anesthesia.[77,78,98,260] Depending on etiology and duration, Laureys et al.[261] showed that cerebral metabolic rates for glucose (CMRGlu) are approximately 40% of normal values, whereas in patients in coma of hypoxic and traumatic origin, values are approximately 50% of normal.[255]

Nonetheless, in some patients residual islands of preserved metabolism have been reported, correlated with fragmentary behavior, such as speaking single words; coordinated, nonpurposeful, nondystonic movements in arms or legs; and strong emotional negativity without motor responses to noxious stimuli with occasional quieting in response to prosodic stimuli. Schiff et al.,[64,65] using [^{18}F]fluorodeoxyglucose-PET (FDG-PET), MRI, and magnetoencephalographic (MEG) responses to sensory stimulation, documented the remaining modular functioning of individual functional networks in PVS cases, although their global metabolic rates remained <50% of normal. They stated that those specific patterns of preserved metabolic activity do not represent random survival of a few neuronal islands, but correspond to the modular nature of individual functional networks that lie beneath a conscious behavior, and that the correlation of different fragmentary behavioral features with preserved metabolic and physiologic activity in cortical and subcortical regions supporting specific modular functions demonstrates the capacity of severely damaged brains to partially express surviving modular functions without evidence of integrative processes that would be necessary to produce consciousness. They concluded that the meager expression of isolated neuropsychologic activity by isolated modules is not enough to generate consciousness in devastatingly damaged brains.

Laureys et al.[78,262] showed context-dependent higher-order auditory processing. They previously reported that in the VS, auditory stimulation activates primary auditory cortices but not higher-order associative areas from which they are disconnected. Their work revealed that functional connectivity between primary and higher-order associative cortices areas can be partially preserved in MCS cases, although disconnected in PVS cases. In this sense, Boly et al.[263] reported similar results exploring cerebral blood flow in PVS and MCS patients to auditory click stimuli by ^{15}O-radiolabeled water–PET. They documented that in MCS patients

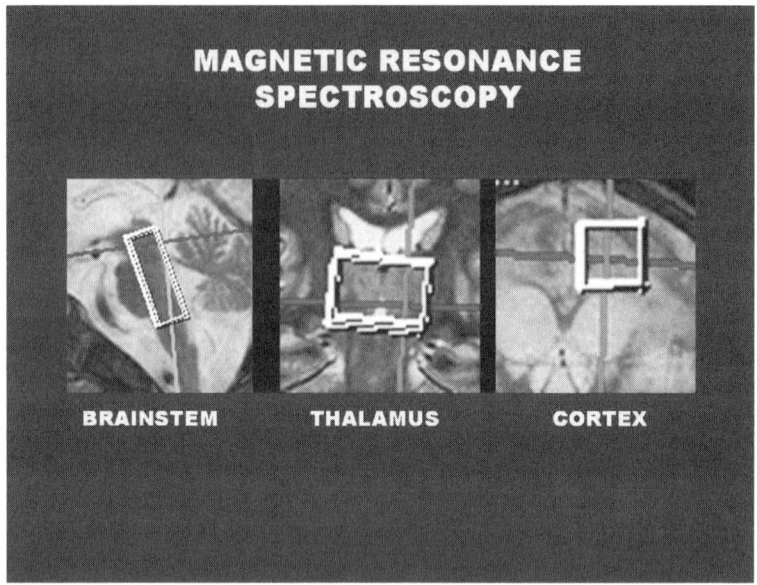

FIGURE 7.3. Regions selected to study magnetic resonance spectroscopy: brainstem, thalamus, and cortex. (See also color insert.)

and in healthy controls, auditory stimulation activated bilateral superior temporal gyri (Brodmann areas 41, 42, and 22); in PVS patients the activation was restricted to Brodmann areas 41 and 42 bilaterally. Boly et al. also concluded that cerebral activity observed in MCS patients is more likely to lead to higher-order integrative processes, although in PVS cases auditory stimulation only activates primary auditory cortices.

Schiff et al.[64,65] recently studied two MCS patients using fMRI to explore cortical responses to passive language. By applying auditory stimulation with personalized narratives, they elicited cortical activity in the superior and middle temporal in both MCS patients, similar to what was found in healthy volunteers. These authors stated that some MCS patients may retain widely distributed cortical systems with the potential for cognitive and sensory function despite their inability to follow simple instructions or communicate reliably, and that the networks remain largely intact but show significant differences in their level of responsiveness.

We recently began a protocol to study PVS and MCS using magnetic resonance spectroscopy (MRS), a technique that provides in-vivo functional analysis of brain metabolites (Machado et al., in preparation). Figures 7.3 and 7.4 show how N-acetylaspartate (NAA) presents a lower concentration in the cerebral cortex, compared to brainstem and thalamus, in a PVS patient. Of course, these are only preliminary results and we need to collect an adequate number of VS and

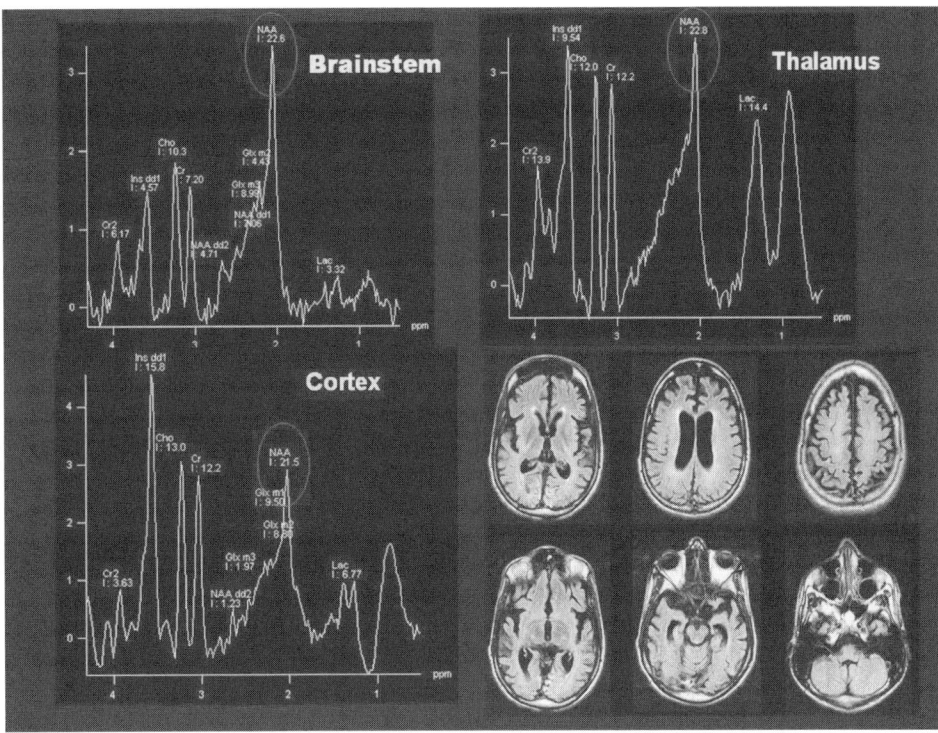

FIGURE 7.4. Magnetic resonance spectroscopy in a PVS case. *N*-acetylaspartate (NAA) presents a lower concentration in cerebral cortex, compared to brainstem and thalamus, in a PVS patient. (See also color insert.)

MCS cases with different pathological findings. Nonetheless, MRS seems to be a useful technique to study in vivo the brains in these cases.

Neurophysiological Techniques

Neurophysiologic techniques have also provided important information about the possibility of remaining cortical processing in PVS cases. Earlier studies using event-related potentials (ERPs) in PVS cases have reported indices of cortical processing.[263–268] Connolly et al.[269] reported a patient with a severe brain injury, in whom unexpectedly positive ERPs findings were later confirmed by this patient's regaining cognitive functions.

Kotchoubey et al.[270] found that among patients with prevailing theta or slow alpha EEG background activity, all patients diagnosed as PVS exhibit some cortical responsivity. In some of them, only primary cortical components N1 or P2 were reliably proven, but more complex cortical responses were also present in these patients. In contrast to other studies using other neuroimaging techniques, these authors documented a percentage of PVS patients having a P3 or P600, indicating activity in association cortical areas.[270] Kotchoubey et al. stated that their results indicate that a high frequency of cortical responses in PVS patients might be due to possible diagnostic errors, or might be evidence that isolated thalamo-cortical circuits are working in PVS patients, indicating spared function of some specialized processing modules, although this processing is unrelated to conscious experience.[60,270–275]

Nonetheless, it is important to consider the time and space resolutions in assessing brain function comparing neurophysiological with other neuroimaging methods. It is well known that ERP and other neurophysiological techniques have a high time resolution, which allows assessing early brain functional changes, in the time dominion of milliseconds, in contrast to cerebral blood flow (CBF) measurements achieved by fMRI, PET, and SPECT, which only express metabolic and functional changes in the brain in seconds and minutes.[276–282] Hence, those results reported by Kotchoubey et al. and other authors, indicating a high frequency of responses in association cortical areas in PVS patients, should be taken into consideration for future research, because these neurophysiological techniques[279,283–289] are capable of providing earlier and different brain functioning signs compared to fMRI, PET, and SPECT.[68,69,277]

Moreover, the lower spatial resolution of neurophysiologic methods may be resolved by calculating ERPs with quantitative electric tomography, which integrates the high time resolution of electrophysiologic measurements and the spatial–anatomical information provided by magnetic resonance images. This technique localizes the generator sources of surface EEG potentials within the brain, based on true anatomy. Hence, ERPs provide a valuable method to assess patients with impaired consciousness.[68,69]

We recently have begun a protocol for assessing awareness in PVS and MCS patients using quantitative electric tomography (QEEGt), calculated with the individual brain studied by anatomic MRI (slices of 1 mm in axial, coronal and saggital planes). The technique for QEEGt is described elsewhere.[300] In Figure 7.5 of we present results from a PVS case. We studied the patient in 10 separate sessions, with an interval of 7 days. Previous to the QEEGt study, clinical examinations demonstrated no changes in his clinical condition. QEEGt was calculated in two conditions: basal record (without any stimulus), and listening to his mothers voice, previously recorded. A grand average of the EEG spectrum was calculated for each condition. A t-Student test was applied to assess mean differences for 49 different EEG frequencies, beginning at 0.78 Hz up to 70 Hz, with intervals of 0.49 Hz. QEEGt maps in an axial plane (A), and in a 3D reconstruction, showed statistical differences (red and yellow colors) comparing both experimental conditions (mother talks vs. basal record) for an EEG frequency of 33. 2 HZ (gamma band), in the posterior and lateral regions of the left hemisphere. This suggests EEG activation when his mother talked to him in these brain areas.

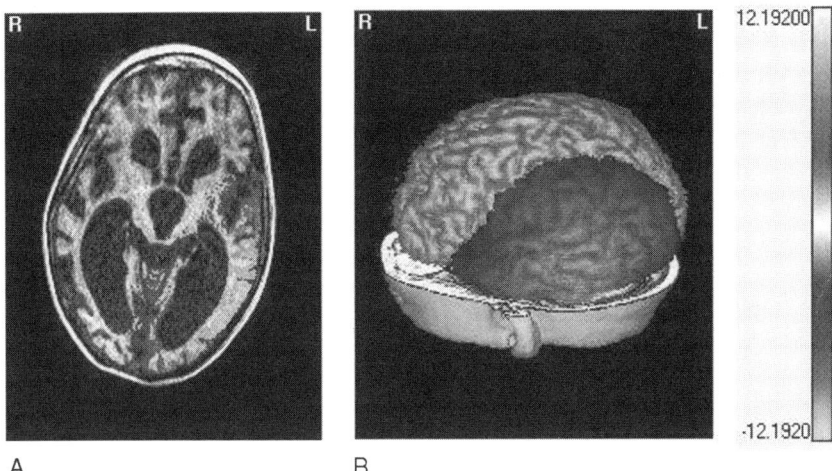

FIGURE 7.5. Quantitative electroencephalographic tomography (QEEGt) is calculated in a PVS patient. QEEGt maps in an axial plane (A) and in a 3D reconstruction. Red and yellow colors show brain regions that might be activated when patient's mother talks to him. (See also color insert.)

We discuss that taking in consideration previous studies of gamma binding and consciousness generation in human beings,[132] these results might demonstrate preservation of awareness in this PVS patient. (Machado et al., in preparation.)

Recuperation in the Persistent Vegetative State

The recent description of the MCS syndrome has demonstrated partial recovery in patients who earlier were in PVS. Laureys et al.[78] stated: "MCS may be a transitional state on the route to further recovery, just like the patient's vegetative state was a transitional state earlier in his course."

Recent studies have reported preserved cortical modular network functioning in five VS patients. One of these patients, who was vegetative for 20 years, spoke single words.[64,65]

Kotchoubey and his group recently documented a case of late recovery of the content of consciousness. The case was a young male patient who was fully vegetative for 20 months, but during a 2-month transition stage this condition converted into an anterograde amnesia; that is, the patient remained severely disabled regarding new learning, but he was completely aware of his person, family, faith, and preaccident biography. Beginning from the sixth month, the authors observed an improvement in his ERPs to complex sensory and verbal stimulation, although the clinical examinations remained unchanged. In the third ERP examination a clear P300 was recorded during an oddball task as well as a differentiated response in the word pair task, indicating cortical processing of word meaning. As Koutcheby et al. have emphasized, recovery of the content of consciousness after 20 months in a VS is an extremely rare event. Therefore, this condition is usually defined as permanent VS, which implies that no regaining of consciousness is possible any longer.

This notion of irreversibility has obvious medical, legal, and ethical implications. Therefore, each well-documented case of such a late awakening should be published.

The development of rehabilitation techniques for VS patients and others suffering long-lasting effects of brain injury is a crucial challenge for neuroscientists. The multisensory stimulation approach, the use of deep brain stimulation, and the implementation of prosthetics to treat cognitive disabilities have provided new hope for cognitive rehabilitation of VS patients.[3,5] Beyond these advances, the use of neural stem cells is a potentially promising method for brain repair.

Do Vegetative State Patients Suffer?

The perceptions of pain and suffering are conscious experiences; unconsciousness, by definition, precludes these experiences.[290,291] The Multi-Society Task Force on PVS concluded that PVS patients are unconscious, and they "cannot experience pain and suffering."[87,88] However, it is important to argue that the pain response in newborns does *not* involve the cerebral cortex, which is one of the primary loci of damage in PVS.[292] It logically follows that PVS patients have the same potential to experience pain and suffering as newborns. This argument also implies that brute animals cannot experience pain since they are not self-conscious. Howsepian[293] considered this to be "at best counterintuitive and at worst patently false."

Kassubek et al.[294] reported on long-term VS assessed by ^{15}O-H$_{(2)}$O PET and residual pain-related cerebral network activity. They reported a pain-induced activation (hyperperfusion) in the posterior insula/secondary somatosensory cortex (SII), postcentral gyrus/primary somatosensory cortex (SI), and the cingulate cortex contralateral to the stimulus and in the posterior insula ipsilateral to the stimulus. No other areas showing complex pain-processing were significantly activated. They concluded that residual cortical pain-related cerebral network could remain in long-term PVS patients.

Laureys et al.[61,62,261] have concluded that high-intensity electrical stimulation produced a robust poststimulus activation in brainstem, thalamus, and primary somatosensory cortex in vegetative patients. Nonetheless, these authors emphasized that higher association areas, such as secondary somatosensory, insular, posterior parietal, and anterior cingulate cortices, were not activated, and they also found that the activated primary somatosensory cortex was isolated from the frontoparietal network, which is thought to be required for conscious perception.

However, Kotchoubey, using ERP, concluded that in a subpopulation of PVS patients with preserved thalamocortical feedback connections, remaining cortical information processing was a consistent finding and may even involve semantic levels of processing. Katayama et al.[295] also reported a persistent increment in pain-related P250, which indicates nonspecific cortical activation.

Nonetheless, as I already discussed, it is important to take into consideration time and space resolution differences between neurophysiologic techniques and other neuroimaging methods. Event-related potentials and other neurophysiologic techniques have a high time resolution, which allows assessing early brain functional changes, in the time dominion of milliseconds, in contrast to CBF measurements achieved by fMRI, PET, and SPECT, which only express metabolic and functional changes in the brain in seconds and minutes.

Hence, can we conversely state that there is no pain if we cannot prove the absence of internal awareness in PVS patients? As Bernat has emphasized, "It is biologically impossible to experience another person's conscious awareness first-hand."

References

1. McHugh P. Annihilating Terri Schiavo. *Hum Life Rev.* 2005;31:67–77.
2. Becker-Schwarze K. Terri Schiavo: the assessment of the persistent vegetative state in accordance with German law. *Eur J Health Law.* 2005;12:321–334.
3. Dute J. The Terri Schiavo case in a comparative perspective. *Eur J Health Law.* 2005;12:317–319.
4. Hentoff N. The legacy of Terri Schiavo: the disabled sound the alarm for the nonreligious. *Free Inq.* 2005;25:33–35.
5. Blendon RJ, Benson JM, Herrmann MJ. The American public and the Terri Schiavo case. *Arch Intern Med.* 2005;165:2580–2584.
6. Rosner F. The Terri Schiavo case in Jewish law. *Cancer Invest.* 2005;23:652.
7. Silverman HJ. Withdrawal of feeding-tubes from incompetent patients: the Terri Schiavo case raises new issues regarding who decides in end-of-life decision making. *Intensive Care Med.* 2005;31:480–481.
8. Johnson M. Terri Schiavo: a disability rights case. *Death Stud.* 2006;30:163–176.
9. Roscoe LA, Osman H, Haley WE. Implications of the Schiavo case for understanding family caregiving issues at the end of life. *Death Stud.* 2006;30:149–161.
10. Preston T, Kelly M. A medical ethics assessment of the case of Terri Schiavo. *Death Stud.* 2006;30:121–133.
11. Wolfson J. Defined by her dying, not her death: the guardian ad litem's view of Schiavo. *Death Stud.* 2006;30:113–120.
12. Cerminara KL. Theresa Marie Schiavo's long road to peace. *Death Stud.* 2006;30:101–112.
13. Saunders WL. Washington insider. *Natl Cathol Bioeth Q.* 2005;5:229–239.
14. Darr K. Terri Schindler Schiavo: end-game. *Hosp Top.* 2005;83:29–31.
15. Goodnough A. U.S. judge denies feeding-tube bid in Schiavo's case; parents quickly appeal; family cites urgency as woman enters 5th day without nutrition. *New York Times.* 2005;A1, A14, April 10, 2005.
16. Goodnough A. U.S. judge hears tense testimony in Schiavo's case; no quick ruling is made; he expresses doubts that a federal review will change outcome. *New York Times.* 2005;A1, A14.

17. Goodnough A. Feed-tube law is struck down in Florida case. *New York Times.* 2004;A1, A20.
18. Newman M. Gov. Bush's role is ended in feeding tube dispute. *New York Times.* 2005;A21, January 25, 2005.
19. Whelan CM. Karen Ann Quinlan: patient or prisoner? *America (NY).* 1975;133:346–347.
20. Battelle P. "Let me sleep": the story of Karen Ann Quinlan. *Ladies Home J.* September 1976 Sep;93(9):69–76.
21. Casey LB. A statement on the case of Karen Ann Quinlan. *Cathol Mind.* 1976;74:12–18.
22. Rome HP. More about Karen Ann Quinlan. *Psychiatr Ann.* 1976;6:97–99.
23. Kohl M. On death, dying, and the Karen Ann Quinlan case. *Humanist.* 1976;36:16.
24. Karen Ann Quinlan, moved, to get 'normal nutrients'. *Washington Post.* 1976;A10, June 11, 1976.
25. Leiman SZ. The Karen Ann Quinlan case: a Jewish perspective. *Gratz Coll Annual Jew Stud.* 1977;6:43–50.
26. Healey JM. The tragic case of Karen Ann Quinlan. *Conn Med.* 1979;43:399.
27. Murphy P. The legacy of Karen Ann Quinlan: a nurses' perspective. *Trends Health Care Law Ethics.* 1993;8:78–79.
28. Kinney HC, Korein J, Panigrahy A, Dikkes P, Goode R. Neuropathological findings in the brain of Karen Ann Quinlan. The role of the thalamus in the persistent vegetative state. *N Engl J Med.* 1994;330:1469–1475.
29. Jeret JS. The brain of Karen Ann Quinlan. *N Engl J Med.* 1994;331:1379–1380.
30. Jellinger KA. The brain of Karen Ann Quinlan. *N Engl J Med.* 1994;331:1378–1379.
31. Kenny RW. A cycle of terms implicit in the idea of medicine: Karen Ann Quinlan as a rhetorical icon and the transvaluation of the ethics of euthanasia. *Health Commun.* 2005;17:17–39.
32. McCormick RA. The Nancy Cruzan case. *Midwest Med Ethics.* 1989;5:2–31.
33. USCC brief in Nancy Cruzan case: continued nutrition and hydration urged. *Origins.* 1989;19:345–351.
34. Veatch RM. Nancy Cruzan and the best interest standard. *Midwest Med Ethics.* 1990;6:17–19.
35. Schwartz RL. Euthanasia and the right to die: Nancy Cruzan and New Mexico. *N M Law Rev.* 1990;20:675–700.
36. Cruzan v. Director, Missouri Department of Health. *Wests Supreme Court Report.* 1990;110:2841–2892.
37. Lewin T. Nancy Cruzan dies, outlived by a debate over the right to die. *New York Times* 1990;A1, A15 December 27, 1990.
38. Singer PA. The case of Nancy Cruzan, the Patient Self-Determination Act and advance directives in Canada. *Humane Med.* 1991;7:225–227.
39. Goldberg S. The changing face of death: computers, consciousness, and Nancy Cruzan. *Stanford Law Rev.* 1991;43:659–684.
40. Attorney for Cruzans discusses legal 'odyssey'. *Iowa Med.* 1991;81:249–250.
41. Cruzan J, White CC. The Cruzans talk about Nancy, the critical care experience and their new mission. Interview by Michael Villaire. *Crit Care Nurse.* 1992;12:80–87.
42. Keyser PK. After Cruzan: the "values base" to advance directives. *Orthop Nurs.* 1992;11:37–40.

43. Doukas DJ, Brody H. After the Cruzan case: the primary care physician and the use of advance directives. *J Am Board Fam Pract.* 1992;5:201–205.

44. Keown J. Courting euthanasia?: Tony Bland and the Law Lords. *Ethics Med.* 1993;9:34–37.

45. Dyer C. Law Lords rule that Tony Bland does not create precedent. *BMJ.* 1993;306:413–414.

46. Dyer C. Tony Bland awaits final hearing. *BMJ.* 1993;306:11–12.

47. Howe JG. Management of patients in persistent vegetative state. Tony Bland had intensive treatment. *BMJ.* 1993;307:625.

48. Michelson DJ, Ashwal S. Evaluation of coma and brain death. *Semin Pediatr Neurol.* 2004;11:105–118.

49. Ashwal S. Recovery of consciousness and life expectancy of children in a vegetative state. *Neuropsychol Rehabil.* 2005;15:190–197.

50. Giacino JT. Disorders of consciousness: differential diagnosis and neuropathologic features. *Semin Neurol.* 1997;17:105–111.

51. Giacino JT, Kalmar K, Whyte J. The JFK Coma Recovery Scale-Revised: measurement characteristics and diagnostic utility. *Arch Phys Med Rehabil.* 2004;85:2020–2029.

52. Kalmar K, Giacino JT. The JFK Coma Recovery Scale–Revised. *Neuropsychol Rehabil.* 2005;15:454–460.

53. Borthwick CJ, Crossley R. Permanent vegetative state: usefulness and limits of a prognostic definition. *NeuroRehabilitation.* 2004;19:381–389.

54. Jennett B, Plum F. Persistent vegetative state after brain damage. A syndrome in search of a name. *Lancet.* 1972;1:734–737.

55. Hassler R, Ore GD, Dieckmann G, Bricolo A, Dolce G. Behavioural and EEG arousal induced by stimulation of unspecific projection systems in a patient with post-traumatic apallic syndrome. *Electroencephalogr Clin Neurophysiol.* 1969;27:306–310.

56. Biniek R, Ferbert A, Rimpel J, et al. The complete apallic syndrome—a case report. *Intensive Care Med.* 1989;15:212–215.

57. Hagel K, Rietz S. [Prognosis of the apallic syndrome. A literature review]. *Anaesthesist.* 1998;47:677–682.

58. Tsymbaliuk VI, Pichkur LD, Martyniuk VI. [Neurotransplantation in the treatment of the sequelae of the apallic syndrome]. *Zh Vopr Neirokhir Im N N Burdenko.* 1998;15–17.

59. Stepan C, Haidinger G, Binder H. Prevalence of persistent vegetative state/apallic syndrome in Vienna. *Eur J Neurol.* 2004;11:461–466.

60. Kotchoubey B. Apallic syndrome is not apallic: is vegetative state vegetative? *Neuropsychol Rehabil.* 2005;15:333–356.

61. Laureys S, Faymonville ME, Peigneux P, et al. Cortical processing of noxious somatosensory stimuli in the persistent vegetative state. *Neuroimage.* 2002;17:732–741.

62. Laureys S. The neural correlate of (un)awareness: lessons from the vegetative state. *Trends Cogn Sci.* 2005;9:556–559.

63. Schiff ND, Plum F. The role of arousal and "gating" systems in the neurology of impaired consciousness. *J Clin Neurophysiol.* 2000;17:438–452.

64. Schiff ND, Ribary U, Moreno DR, et al. Residual cerebral activity and behavioural fragments can remain in the persistently vegetative brain. *Brain.* 2002;125:1210–1234.

65. Schiff ND, Rodriguez-Moreno D, Kamal A, et al. fMRI reveals large-scale network activation in minimally conscious patients. *Neurology.* 2005;64:514–523.

66. Gonzalez Andino SL, Pascual Marqui RD, Valdes Sosa PA, et al. Brain electrical field measurements unaffected by linked earlobes reference. *Electroencephalogr Clin Neurophysiol.* 1990;75:155–160.
67. Machado C. The minimally conscious state: definition and diagnostic criteria. *Neurology.* 2002;59:1473–1474.
68. Machado C. Cerebral processing in the minimally conscious state. *Neurology.* 2005;65:973–974.
69. Machado C. Can vegetative state patients retain cortical processing? *Clin Neurophysiol.* 2005;116:2253–2254.
70. Bernat JL. Questions remaining about the minimally conscious state. *Neurology.* 2002;58:337–338.
71. Leong B. The vegetative and minimally conscious states in children: spasticity, muscle contracture and issues for physiotherapy treatment. *Brain Inj.* 2002;16:217–230.
72. Giacino JT. The minimally conscious state: defining the borders of consciousness. *Prog Brain Res.* 2005;150:381–395.
73. Giacino J, Whyte J. The vegetative and minimally conscious states: current knowledge and remaining questions. *J Head Trauma Rehabil.* 2005;20:30–50.
74. Munday R. Vegetative and minimally conscious states: how can occupational therapists help? *Neuropsychol Rehabil.* 2005;15:503–513.
75. Elliott L, Walker L. Rehabilitation interventions for vegetative and minimally conscious patients. *Neuropsychol Rehabil.* 2005;15:480–493.
76. Bernat JL, Culver CM, Gert B. On the definition and criterion of death. *Arch Intern Med.* 1981;94:389–394.
77. Faymonville ME, Pantke KH, Berre J, et al. [Cerebral functions in brain-damaged patients. What is meant by coma, vegetative state, minimally conscious state, locked-in syndrome and brain death?]. *Anaesthesist.* 2004;53:1195–1202.
78. Laureys S, Owen AM, Schiff ND. Brain function in coma, vegetative state, and related disorders. *Lancet Neurol.* 2004;3:537–546.
79. Giacino JT. The vegetative and minimally conscious states: consensus-based criteria for establishing diagnosis and prognosis. *NeuroRehabilitation.* 2004;19:293–298.
80. Giacino JT, Kalmar K. Diagnostic and prognostic guidelines for the vegetative and minimally conscious states. *Neuropsychol Rehabil.* 2005;15:166–174.
81. Childs NL, Mercer WN. Misdiagnosing the persistent vegetative state. Misdiagnosis certainly occurs. *BMJ.* 1996;313:944.
82. Wilson BA. Part III: Behavioural assessment and rehabilitation techniques. *Neuropsychol Rehabil.* 2005;15:428–430.
83. Wilson FC, Graham LE, Watson T. Vegetative and minimally conscious states: serial assessment approaches in diagnosis and management. *Neuropsychol Rehabil.* 2005;15:431–441.
84. Stevens RD, Bhardwaj A. Approach to the comatose patient. *Crit Care Med.* 2006;34:31–41.
85. Zingler VC, Pohlmann-Eden B. Diagnostic pitfalls in patients with hypoxic brain damage: three case reports. *Resuscitation.* 2005;65:107–110.
86. Kinney HC, Samuels MA. Neuropathology of the persistent vegetative state. A review. *J Neuropathol Exp Neurol.* 1994;53:548–558.
87. Medical aspects of the persistent vegetative state 2. The Multi-Society Task Force on PVS. *N Engl J Med.* 1994;330:1572–1579.
88. Medical aspects of the persistent vegetative state 1. The Multi-Society Task Force on PVS. *N Engl J Med.* 1994;330:1499–1508.

89. Bernat JL. *Ethical Issues in Neurology*. Philadelphia: Lippincott, 1991.
90. Andrews K. Recovery of patients after four months or more in the persistent vegetative state. *BMJ*. 1993;306:1597–1600.
91. Andrews K, Murphy L, Munday R, Littlewood C. Misdiagnosis of the vegetative state: retrospective study in a rehabilitation unit. *BMJ*. 1996;313:13–16.
92. Andrews K. Rehabilitation practice following profound brain damage. *Neuropsychol Rehabil*. 2005;15:461–472.
93. Malik K, Hess DC. Evaluating the comatose patient. Rapid neurologic assessment is key to appropriate management. *Postgrad Med*. 2002;111:38–36, 49.
94. Ng YS, Chua KS. States of severely altered consciousness: clinical characteristics, medical complications and functional outcome after rehabilitation. *NeuroRehabilitation*. 2005;20:97–105.
95. Ansell BJ, Keenan JE. The Western Neuro Sensory Stimulation Profile: a tool for assessing slow-to-recover head-injured patients. *Arch Phys Med Rehabil*. 1989;70:104–108.
96. Shiel A, Horn SA, Wilson BA, Watson MJ, Campbell MJ, McLellan DL. The Wessex Head Injury Matrix (WHIM) main scale: a preliminary report on a scale to assess and monitor patient recovery after severe head injury. *Clin Rehabil*. 2000;14:408–416.
97. Giacino JT, Ashwal S, Childs N, et al. The minimally conscious state: definition and diagnostic criteria. *Neurology*. 2002;58:349–353.
98. Cohen-Almagor R. Some observations on post-coma unawareness patients and on other forms of unconscious patients: policy proposals. *Med Law*. 1997;16:451–471.
99. Childs NL, Mercer WN, Childs HW. Accuracy of diagnosis of persistent vegetative state. *Neurology*. 1993;43:1465–1467.
100. Childs NL, Childs A. Coma and vegetative states. *Arch Phys Med Rehabil*. 1993;74:333.
101. Childs NL, Mercer WN. Brief report: late improvement in consciousness after post-traumatic vegetative state. *N Engl J Med*. 1996;334:24–25.
102. Childs N. Differences in vegetative states. *Tex Med*. 1999;95:6.
103. Pallis C. Diagnosis of brain death. *Br Med J*. 1980;281:1491–1492.
104. Plum P. Coma and related global disturbances of the human conscious state. In: Peters A, ed. *Cerebral Cortex*. New York: Plenum, 1991:359–425.
105. Johkura K, Komiyama A, Kuroiwa Y. Vertical conjugate eye deviation in postresuscitation coma. *Ann Neurol*. 2004;56:878–881.
106. Baars BJ. Global workspace theory of consciousness: toward a cognitive neuroscience of human experience. *Prog Brain Res*. 2005;150:45–53.
107. Wijdicks EF, Bamlet WR, Maramattom BV, Manno EM, McClelland RL. Validation of a new coma scale: The FOUR score. *Ann Neurol*. 2005;58:585–593.
108. Plum F, Posner JB. *The Diagnosis of Stupor and Coma*. Philadelphia: FA Davis, 1980.
109. Yamamoto T, Kobayashi K, Kasai M, Oshima H, Fukaya C, Katayama Y. DBS therapy for the vegetative state and minimally conscious state. *Acta Neurochir Suppl*. 2005;93:101–104.
110. Cairns H, Oldfield RC, Pennybacker JB, and Whitteridge D. Akinetic mutism with epidermoid cyst of the third ventricle (with a report on associated disturbance of brain potentials). *Brain*. 1941;64:273–290.
111. Wijdicks EF, Cranford RE. Clinical diagnosis of prolonged states of impaired consciousness in adults. *Mayo Clin Proc*. 2005;80:1037–1046.

112. Recommendations for use of uniform nomenclature pertinent to patients with severe alterations in consciousness. American Congress of Rehabilitation Medicine. *Arch Phys Med Rehabil.* 1995;76:205–209.

113. Devinsky O, Morrell MJ, Vogt BA. Contributions of anterior cingulate cortex to behaviour. *Brain.* 1995;118(pt 1):279–306.

114. Buge A, Escourolle R, Rancurel G, Poisson M. [Akinetic mutism and bicingular softening. 3 anatomo-clinical cases]. *Rev Neurol (Paris).* 1975;131:121–131.

115. Castaigne P, Lhermitte F, Buge A, Escourolle R, Hauw JJ, Lyon-Caen O. Paramedian thalamic and midbrain infarct: clinical and neuropathological study. *Ann Neurol.* 1981;10:127–148.

116. Psarros T, Zouros A, Coimbra C. Bromocriptine-responsive akinetic mutism following endoscopy for ventricular neurocysticercosis. Case report and review of the literature. *J Neurosurg.* 2003;99:397–401.

117. Choudhari KA. Subarachnoid haemorrhage and akinetic mutism. *Br J Neurosurg.* 2004;18:253–258.

118. Oberndorfer S, Urbanits S, Lahrmann H, Kirschner H, Kumpan W, Grisold W. Akinetic mutism caused by bilateral infiltration of the fornix in a patient with astrocytoma. *Eur J Neurol.* 2002;9:311–313.

119. Llinas RR, Pare D. Of dreaming and wakefulness. *Neuroscience.* 1991;44:521–535.

120. Machado C. A new definition of death based on the basic mechanisms of consciousness generation in human beings. In: Machado C, ed. *Brain Death (Proceedings of the Second International Symposium on Brain Death).* Amsterdam: Elsevier Science BV, 1995:57–66.

121. Damasio AR. Consciousness. Knowing how, knowing where. *Nature.* 1995;375:106–107.

122. Pare D, Llinas R. Conscious and pre-conscious processes as seen from the standpoint of sleep-waking cycle neurophysiology. *Neuropsychologia.* 1995;33:1155–1168.

123. Bogen JE. On the neurophysiology of consciousness: Part II. Constraining the semantic problem. *Conscious Cogn.* 1995;4:137–158.

124. Bogen JE. On the neurophysiology of consciousness: I. An overview. *Conscious Cogn.* 1995;4:52–62.

125. Koch C. Visual awareness and the thalamic intralaminar nuclei. *Conscious Cogn.* 1995;4:163–166.

126. Lindahl BI. Consciousness and biological evolution. *J Theor Biol.* 1997;187:613–629.

127. Cariani P. Consciousness and the organization of neural processes: commentary on John, et al. *Conscious Cogn.* 1997;6:56–64.

128. Bogen JE. Some neurophysiologic aspects of consciousness. *Semin Neurol.* 1997;17:95–103.

129. Feinberg TE. The irreducible perspectives of consciousness. *Semin Neurol.* 1997;17:85–93.

130. Damasio AR. Investigating the biology of consciousness. *Philos Trans R Soc Lond B Biol Sci.* 1998;353:1879–1882.

131. Tononi G, Edelman GM. Consciousness and complexity. *Science.* 1998;282:1846–1851.

132. Llinas R, Ribary U, Contreras D, Pedroarena C. The neuronal basis for consciousness. *Philos Trans R Soc Lond B Biol Sci.* 1998;353:1841–1849.

133. Llinas RR, Ribary U. Temporal conjunction in thalamocortical transactions. *Adv Neurol.* 1998;77:95–102.

134. Jones EG. A new view of specific and nonspecific thalamocortical connections. *Adv Neurol.* 1998;77:49–71.

135. Jones EG. Viewpoint: the core and matrix of thalamic organization. *Neuroscience.* 1998;85:331–345.

136. Shewmon DA, Holmes GL, Byrne PA. Consciousness in congenitally decorticate children: developmental vegetative state as self-fulfilling prophecy. *Dev Med Child Neurol.* 1999;41:364–374.

137. Machado C. Consciousness as a definition of death: its appeal and complexity. *Clin Electroencephalogr.* 1999;30:156–164.

138. Lloyd D. Beyond "the fringe": A cautionary critique of William James. *Conscious Cogn.* 2000;9:629–637.

139. Fernandez dM. [Thalamo-cortical system and consciousness]. *An R Acad Nac Med (Madr).* 2000;117:855–869.

140. Latronico N, Alongi S, Guarneri B, Cappa S, Candiani A. [Approach to the patient in vegetative state. Part I: diagnosis]. *Minerva Anestesiol.* 2000;66:225–231.

141. Edelman G. Consciousness: the remembered present. *Ann N Y Acad Sci.* 2001;929:111–122.

142. Parvizi J, Damasio A. Consciousness and the brainstem. *Cognition.* 2001;79:135–160.

143. Taylor JG. The central role of the parietal lobes in consciousness. *Conscious Cogn.* 2001;10:379–417.

144. Pribram K. Commentary on E. R. John, et al. "Invariant reversible QEEG effects of anesthetics" and E. R. John "A field theory of consciousness". *Conscious Cogn.* 2001;10:214–216.

145. Llinas R, Ribary U. Consciousness and the brain. The thalamocortical dialogue in health and disease. *Ann N Y Acad Sci.* 2001;929:166–175.

146. Machado-Curbelo C. [Do we defend a brain oriented view of death?]. *Rev Neurol.* 2002;35:387–396.

147. Pastor-Gomez J. [Quantum mechanics and brain: a critical review]. *Rev Neurol.* 2002;35:87–94.

148. John ER. The neurophysics of consciousness. *Brain Res Brain Res Rev.* 2002;39:1–28.

149. Kotchoubey B, Lang S, Bostanov V, Birbaumer N. Is there a mind? Electrophysiology of unconscious patients. *News Physiol Sci.* 2002;17:38–42.

150. Parvizi J, Damasio AR. Neuroanatomical correlates of brainstem coma. *Brain.* 2003;126:1524–1536.

151. Zingler VC, Krumm B, Bertsch T, Fassbender K, Pohlmann-Eden B. Early prediction of neurological outcome after cardiopulmonary resuscitation: a multimodal approach combining neurobiochemical and electrophysiological investigations may provide high prognostic certainty in patients after cardiac arrest. *Eur Neurol.* 2003;49:79–84.

152. Rousselet GA, Thorpe SJ, Fabre-Thorpe M. How parallel is visual processing in the ventral pathway? *Trends Cogn Sci.* 2004;8:363–370.

153. Lizza JP. The conceptual basis for brain death revisited: loss of organic integration or loss of consciousness? *Adv Exp Med Biol.* 2004;550:51–59.

154. Veatch RM. The death of whole-brain death: the plague of the disaggregators, somaticists, and mentalists. *J Med Philos.* 2005;30:353–378.

155. Laureys S, Pellas F, Van EP, et al. The locked-in syndrome: what is it like to be conscious but paralyzed and voiceless? *Prog Brain Res.* 2005;150:495–511.

156. Machado C. Death on neurological grounds. *J Neurosurg Sci.* 1994;38:209–222.

157. Machado C. Is the concept of brain death secure? In: Zeman A, Enamorado A, eds. *Ethical Dilemmas in Neurology*. London: WB Saunders, 2000:193–212.

158. Shewmon DA. The metaphysics of brain death, persistent vegetative state and dementia. *Thomist.* 1985;49:24–80.

159. Ashwal S, Bale JF, Jr., Coulter DL, et al. The persistent vegetative state in children: report of the Child Neurology Society Ethics Committee. *Ann Neurol.* 1992;32:570–576.

160. Shewmon DA. Spinal shock and 'brain death': somatic pathophysiological equivalence and implications for the integrative-unity rationale. *Spinal Cord.* 1999;37:313–324.

161. Shewmon DA, Shewmon ES. The semiotics of death and its medical implications. *Adv Exp Med Biol.* 2004;550:89–114.

162. Destexhe A, Contreras D, Sejnowski TJ, Steriade M. A model of spindle rhythmicity in the isolated thalamic reticular nucleus. *J Neurophysiol.* 1994;72:803–818.

163. Steriade M. Brain activation, then and now: coherent fast rhythms in corticothalamic networks. *Arch Ital Biol.* 1995;134:5–20.

164. Steriade M, Contreras D, Amzica F, Timofeev I. Synchronization of fast (30–40 Hz) spontaneous oscillations in intrathalamic and thalamocortical networks. *J Neurosci.* 1996;16:2788–2808.

165. Bazhenov M, Timofeev I, Steriade M, Sejnowski TJ. Cellular and network models for intrathalamic augmenting responses during 10-Hz stimulation. *J Neurophysiol.* 1998;79:2730–2748.

166. Wassle H. Parallel processing in the mammalian retina. *Nat Rev Neurosci.* 2004;5:747–757.

167. Llinas R, Urbano FJ, Leznik E, Ramirez RR, van Marle HJ. Rhythmic and dysrhythmic thalamocortical dynamics: GABA systems and the edge effect. *Trends Neurosci.* 2005;28:325–333.

168. Fuentealba P, Steriade M. The reticular nucleus revisited: intrinsic and network properties of a thalamic pacemaker. *Prog Neurobiol.* 2005;75:125–141.

169. Koch G, Oliveri M, Torriero S, Carlesimo GA, Turriziani P, Caltagirone C. rTMS evidence of different delay and decision processes in a fronto-parietal neuronal network activated during spatial working memory. *Neuroimage.* 2005;24:34–39.

170. Achard S, Salvador R, Whitcher B, Suckling J, Bullmore E. A resilient, low-frequency, small-world human brain functional network with highly connected association cortical hubs. *J Neurosci.* 2006;26:63–72.

171. Mottaghy FM, Willmes K, Horwitz B, Muller HW, Krause BJ, Sturm W. Systems level modeling of a neuronal network subserving intrinsic alertness. *Neuroimage.* 2006;29:225–233.

172. Kretschmer E. Das apallische syndrom. *Z Ges Neurol Psychiatry.* 1940;169: 576–579.

173. Ingvar DH. The pathophysiology of cerebral anoxia. *Acta Anaesthesiol Scand Suppl.* 1968;29:47–59.

174. Ingvar DH, Brun A. [The complete apallic syndrome]. *Arch Psychiatr Nervenkr.* 1972;215:219–239.

175. Ingvar DH. Clinical neurophysiology of the cerebral circulation. *Electroencephalogr Clin Neurophysiol Suppl.* 1978;71–81.

176. De Volder AG, Michel C, Guérit JM, et al. Brain glucose metabolism in postanoxic syndrome due to cardiac arrest. *Acta Neurol Belg.* 1994;94:183–189.

177. Rothstein TL. Recovery from near death following cerebral anoxia: a case report demonstrating superiority of median somatosensory evoked potentials over EEG in

predicting a favorable outcome after cardiopulmonary resuscitation. *Resuscitation.* 2004;60:335–341.

178. Wilson FC, Harpur J, Watson T, Morrow JI. Adult survivors of severe cerebral hypoxia—case series survey and comparative analysis. *NeuroRehabilitation.* 2003;18:291–298.

179. Cole G, Cowie VA. Long survival after cardiac arrest: case report and neuropathological findings. *Clin Neuropathol.* 1987;6:104–109.

180. Bernat JL. The boundaries of the persistent vegetative state. *J Clin Ethics.* 1992;3:176–180.

181. McMahan J. The metaphysics of brain death. *Bioethics.* 1995;9:91–126.

182. Adams JH, Graham DI, Jennett B. The neuropathology of the vegetative state after an acute brain insult. *Brain.* 2000;123(pt 7):1327–1338.

183. Jennett B, Adams JH, Murray LS, Graham DI. Neuropathology in vegetative and severely disabled patients after head injury. *Neurology.* 2001;56:486–490.

184. Graham DI, Adams JH, Murray LS, Jennett B. Neuropathology of the vegetative state after head injury. *Neuropsychol Rehabil.* 2005;15:198–213.

185. Ingvar DH. EEG and cerebral circulation in the apallic syndrome and akinetic mutism. *Electroencephalogr Clin Neurophysiol.* 1971;30:272–273.

186. Ingvar DH, Brun A, Johansson L, Sammuelsson SM. Survival after severe cerebral anoxia with destruction of the cerebral cortex: the apallic syndrome. In: Korein J, ed. *Brain Death: Interrelated Medical and Social Issues.* New York: Ann NY Acad Sci, 1977:184–214.

187. Walton DN. Neocortical versus whole-brain conceptions of personal death. *Omega (Westport).* 1981;12:339–344.

188. Puccetti R. Does anyone survive neocortical death? In: Zaner RM, ed. *Death: Beyond Whole Brain Criteria.* Boston: Kluwer Academics, 1988:75–90.

189. Herreras O, Largo C. [Tracking along the path toward ischemic neuronal death]. *Rev Neurol.* 2002;35:838–845.

190. Scorziello A, Pellegrini C, Forte L, et al. Differential vulnerability of cortical and cerebellar neurons in primary culture to oxygen glucose deprivation followed by reoxygenation. *J Neurosci Res.* 2001;63:20–26.

191. Ingvar DH. Brain death—total brain infarction. *Acta Anaesthesiol Scand Suppl.* 1971;45:129–140.

192. Tagami M, Yamagata K, Ikeda K, et al. Genetic vulnerability of cortical neurons isolated from stroke-prone spontaneously hypertensive rats in hypoxia and oxygen reperfusion. *Hypertens Res.* 1999;22:23–29.

193. Kadar T, Dachir S, Shukitt-Hale B, Levy A. Sub-regional hippocampal vulnerability in various animal models leading to cognitive dysfunction. *J Neural Transm.* 1998;105:987–1004.

194. Petito CK, Olarte JP, Roberts B, Nowak TS, Jr., Pulsinelli WA. Selective glial vulnerability following transient global ischemia in rat brain. *J Neuropathol Exp Neurol.* 1998;57:231–238.

195. McDonald JW, Althomsons SP, Hyrc KL, Choi DW, Goldberg MP. Oligodendrocytes from forebrain are highly vulnerable to AMPA/kainate receptor-mediated excitotoxicity. *Nat Med.* 1998;4:291–297.

196. Rap ZM. Influence of damage of hypothalamo-hypophysial neurosecretory system and adrenalectomy on the blood-brain barrier alteration during intracranial hypertension. *Acta Neuropathol Suppl (Berl).* 1981;7:10–12.

197. Geocadin RG, Muthuswamy J, Sherman DL, Thakor NV, Hanley DF. Early electrophysiological and histologic changes after global cerebral ischemia in rats. *Mov Disord.* 2000;15(suppl 1):14–21.
198. Villablanca J, Salinas-Zeballos ME. Sleep-wakefulness, EEG and behavioral studies of chronic cats without the thalamus: the 'athalamic' cat. *Arch Ital Biol.* 1972;110:383–411.
199. Graham DI, McLellan D, Adams JH, Doyle D, Kerr A, Murray LS. The neuropathology of the vegetative state and severe disability after non-missile head injury. *Acta Neurochir Suppl (Wien).* 1983;32:65–67.
200. Cordobes F, Lobato RD, Rivas JJ, et al. Post-traumatic diffuse axonal brain injury. Analysis of 78 patients studied with computed tomography. *Acta Neurochir (Wien).* 1986;81:27–35.
201. Parker JR, Parker JC Jr, Overman JC. Intracranial diffuse axonal injury at autopsy. *Ann Clin Lab Sci.* 1990;20:220–224.
202. Yokota H, Henmi H, Mashiko K, et al. [Clinical study of intracranial pressure and auditory brain stem response in the cases of diffuse axonal injury]. *No Shinkei Geka.* 1992;20:217–221.
203. Lang DA, Teasdale GM, Macpherson P, Lawrence A. Diffuse brain swelling after head injury: more often malignant in adults than children? *J Neurosurg.* 1994;80:675–680.
204. Heath DL, Vink R. Impact acceleration-induced severe diffuse axonal injury in rats: characterization of phosphate metabolism and neurologic outcome. *J Neurotrauma.* 1995;12:1027–1034.
205. Vuilleumier P, Assal G. [Lesions of the corpus callosum and syndromes of interhemispheric disconnection of traumatic origin]. *Neurochirurgie.* 1995;41:98–107.
206. Reider-Groswasser I, Costeff H, Sazbon L, Groswasser Z. CT findings in persistent vegetative state following blunt traumatic brain injury. *Brain Inj.* 1997;11:865–870.
207. Chiaretti A, Visocchi M, Viola L, et al. [Diffuse axonal lesions in childhood]. *Pediatr Med Chir.* 1998;20:393–397.
208. Wang H, Duan G, Zhang J, Zhou D. Clinical studies on diffuse axonal injury in patients with severe closed head injury. *Chin Med J (Engl).* 1998;111:59–62.
209. Kampfl A, Franz G, Aichner F, et al. The persistent vegetative state after closed head injury: clinical and magnetic resonance imaging findings in 42 patients. *J Neurosurg.* 1998;88:809–816.
210. Adams JH, Jennett B, McLellan DR, Murray LS, Graham DI. The neuropathology of the vegetative state after head injury. *J Clin Pathol.* 1999;52:804–806.
211. Gentleman SM, McKenzie JE, Royston MC, McIntosh TK, Graham DI. A comparison of manual and semi-automated methods in the assessment of axonal injury. *Neuropathol Appl Neurobiol.* 1999;25:41–47.
212. Giacino JT, Whyte J. Amantadine to improve neurorecovery in traumatic brain injury-associated diffuse axonal injury: a pilot double-blind randomized trial. *J Head Trauma Rehabil.* 2003;18:4–5.
213. Tong KA, Ashwal S, Holshouser BA, et al. Diffuse axonal injury in children: clinical correlation with hemorrhagic lesions. *Ann Neurol.* 2004;56:36–50.
214. Ginsberg MD, Hedley-Whyte ET, Richardson EP, Jr. Hypoxic-ischemic leukoencephalopathy in man. *Trans Am Neurol Assoc.* 1974;99:213–216.
215. Ginsberg MD, Hedley-Whyte ET, Richardson EP, Jr. Hypoxic-ischemic leukoencephalopathy in man. *Arch Neurol.* 1976;33:5–14.
216. Heye N, Terstegge K, Gosztonyi G. Vasculitic, hypoxic-ischemic leukoencephalopathy. *Clin Neurol Neurosurg.* 1994;96:156–160.

217. Roman GC. Vascular dementia today. *Rev Neurol (Paris)*. 1999;155(suppl 4):S64–S72.

218. Graham DI, Maxwell WL, Adams JH, Jennett B. Novel aspects of the neuropathology of the vegetative state after blunt head injury. *Prog Brain Res*. 2005;150:445–455.

219. Tsubokawa T, Yamamoto T, Katayama Y, Hirayama T, Maejima S, Moriya T. Deep-brain stimulation in a persistent vegetative state: follow-up results and criteria for selection of candidates. *Brain Inj*. 1990;4:315–327.

220. Pappata S, Mazoyer B, Tran DS, Cambon H, Levasseur M, Baron JC. Effects of capsular or thalamic stroke on metabolism in the cortex and cerebellum: a positron tomography study. *Stroke*. 1990;21:519–524.

221. Steriade M. Synchronized activities of coupled oscillators in the cerebral cortex and thalamus at different levels of vigilance. *Cereb Cortex*. 1997;7(6):583–604.

222. Savage LM, Castillo R, Langlais PJ. Effects of lesions of thalamic intralaminar and midline nuclei and internal medullary lamina on spatial memory and object discrimination. *Behav Neurosci*. 1998;112:1339–1352.

223. Jeljeli M, Strazielle C, Caston J, Lalonde R. Effects of centrolateral or medial thalamic lesions on motor coordination and spatial orientation in rats. *Neurosci Res*. 2000;38:155–164.

224. Laureys S, Faymonville ME, Luxen A, Lamy M, Franck G, Maquet P. Restoration of thalamocortical connectivity after recovery from persistent vegetative state. *Lancet*. 2000;355:1790–1791.

225. Maxwell WL, Pennington K, MacKinnon MA, et al. Differential responses in three thalamic nuclei in moderately disabled, severely disabled and vegetative patients after blunt head injury. *Brain*. 2004;127:2470–2478.

226. Mitchell AS, Dalrymple-Alford JC. Dissociable memory effects after medial thalamus lesions in the rat. *Eur J Neurosci*. 2005;22:973–985.

227. Sherman SM. Thalamic relays and cortical functioning. *Prog Brain Res*. 2005;149:107–126.

228. Sugiyama K, Muteki T, Shimoji K. Halothane-induced hyperpolarization and depression of postsynaptic potentials of guinea pig thalamic neurons in vitro. *Brain Res*. 1992;576:97–103.

229. Kinsbourne M. The intralaminar thalamic nuclei: subjectivity pumps or attention-action coordinators? *Conscious Cogn*. 1995;4:167–171.

230. McAlonan K, Brown VJ. The thalamic reticular nucleus: more than a sensory nucleus? *Neuroscientist*. 2002;8:302–305.

231. Rondot P, de Recondo J, Davous P, Bathien N, Coignet A. [Bilateral thalamic infarcts with abnormal movements and permanent amnesia]. *Rev Neurol (Paris)*. 1986;142:398–405.

232. Karacostas D, Artemis N, Giannopoulos S, Milonas I, Bogousslavsky J. Bilateral thalamic infarcts presenting as acute pseudobulbar palsy. *Funct Neurol*. 1994;9:265–268.

233. Fukatsu R, Fujii T, Yamadori A, Nagasawa H, Sakurai Y. Persisting childish behavior after bilateral thalamic infarcts. *Eur Neurol*. 1997;37:230–235.

234. Chalela J, Raps E. Single thalamic-subthalamic artery and bilateral thalamic infarcts. *J Neurol Neurosurg Psychiatry*. 1999;67:119.

235. Goldman-Rakic PS. The physiological approach: functional architecture of working memory and disordered cognition in schizophrenia. *Biol Psychiatry*. 1999;46:650–661.

236. Paul RH, Cohen RA, Moser DJ, et al. Clinical correlates of cognitive decline in vascular dementia. *Cogn Behav Neurol*. 2003;16:40–46.

237. Jellinger KA. Pathology and pathophysiology of vascular cognitive impairment. A critical update. *Panminerva Med.* 2004;46:217–226.
238. Lagares A, de Toledo M, Gonzalez-Leon P, et al. [Bilateral thalamic gliomas: report of a case with cognitive impairment]. *Rev Neurol.* 2004;38:244–246.
239. Bastos Leite AJ, van Straaten EC, Scheltens P, Lycklama G, Barkhof F. Thalamic lesions in vascular dementia: low sensitivity of fluid-attenuated inversion recovery (FLAIR) imaging. *Stroke.* 2004;35:415–419.
240. Potts R, Leech RW. Thalamic dementia: an example of primary astroglial dystrophy of Seitelberger. *Clin Neuropathol.* 2005;24:271–275.
241. Gold G, Kovari E, Herrmann FR, et al. Cognitive consequences of thalamic, basal ganglia, and deep white matter lacunes in brain aging and dementia. *Stroke.* 2005;36:1184–1188.
242. Jellinger KA. Understanding the pathology of vascular cognitive impairment. *J Neurol Sci.* 2005;229–230:57–63.
243. Laureys S, Antoine S, Boly M, et al. Brain function in the vegetative state. *Acta Neurol Belg.* 2002;102:177–185.
244. Fins JJ, Schiff ND. The afterlife of Terri Schiavo. *Hastings Cent Rep.* 2005;35:8.
245. Peterson BS, Skudlarski P, Gatenby JC, Zhang H, Anderson AW, Gore JC. An fMRI study of Stroop word-color interference: evidence for cingulate subregions subserving multiple distributed attentional systems. *Biol Psychiatry.* 1999;45:1237–1258.
246. Hinterberger T, Veit R, Strehl U, et al. Brain areas activated in fMRI during self-regulation of slow cortical potentials (SCPs). *Exp Brain Res.* 2003;152:113–122.
247. Abdel-Dayem HM, Bahar RH, Sigurdsson GH, Sadek S, Olivecrona H, Ali AM. The hollow skull: a sign of brain death in Tc-99m HM-PAO brain scintigraphy. *Clin Nucl Med.* 1989;14:912–916.
248. Yoshikai T, Tahara T, Kuroiwa T, et al. Plain CT findings of brain death confirmed by hollow skull sign in brain perfusion SPECT. *Radiat Med.* 1997;15:419–424.
249. Tsuchida T, Sadato N, Nishizawa S, et al. [99mTc-HMPAO SPECT in the brain death—a case report]. *Kaku Igaku.* 1993;30:663–667.
250. Bonetti MG, Ciritella P, Valle G, Perrone E. 99mTc HM-PAO brain perfusion SPECT in brain death. *Neuroradiology.* 1995;37:365–369.
251. Kahveci F, Bekar A, Tamgac F. Tc-99 HMPAO cerebral SPECT imaging in brain death patients with complex spinal automatism. *Ulus Travma Derg.* 2002;8:198–201.
252. Okuyaz C, Gucuyener K, Karabacak NI, Aydin K, Serdaroglu A, Cingi E. Tc-99m-HMPAO SPECT in the diagnosis of brain death in children. *Pediatr Int.* 2004;46:711–714.
253. Al-Shammri S, Al-Feeli M. Confirmation of brain death using brain radionuclide perfusion imaging technique. *Med Princ Pract.* 2004;13:267–272.
254. Momose T, Nishikawa J, Watanabe T, et al. [Clinical application of 18F-FDG-PET in patients with brain death]. *Kaku Igaku.* 1992;29:1139–1142.
255. Tommasino C, Grana C, Lucignani G, Beretta L, Torri G, Fazio F. [Cerebral metabolism with PET methods in patients in coma and in postcomatose syndrome. A prognostic index?]. *Minerva Anestesiol.* 1993;59:837–841.
256. Vander Borght T, Laloux P, Maes A, Salmon E, Goethals I, Goldman S. Guidelines for brain radionuclide imaging. Perfusion single photon computed tomography (SPECT) using Tc-99m radiopharmaceuticals and brain metabolism positron

emission tomography (PET) using F-18 fluorodeoxyglucose. The Belgian Society for Nuclear Medicine. *Acta Neurol Belg.* 2001;101:196–209.

257. Baldy-Moulinier M, Billiard M, Passouant P. [Simultaneous study of blood flow and cerebral metabolism in prolonged coma]. *Lille Med.* 1976;21:563–566.

258. Agardh CD, Rosen I. Neurophysiological recovery after hypoglycemic coma in the rat: correlation with cerebral metabolism. *J Cereb Blood Flow Metab.* 1983;3:78–85.

259. Tenjin H, Ueda S, Mizukawa N, et al. Positron emission tomographic studies on cerebral hemodynamics in patients with cerebral contusion. *Neurosurgery.* 1990;26:971–979.

260. Juengling FD, Kassubek J, Huppertz HJ, Krause T, Els T. Separating functional and structural damage in persistent vegetative state using combined voxel-based analysis of 3–D MRI and FDG-PET. *J Neurol Sci.* 2005;228:179–184.

261. Laureys S, Faymonville ME, de Tiege X, et al. Brain function in the vegetative state. *Adv Exp Med Biol.* 2004;550:229–238.

262. Boly M, Faymonville ME, Peigneux P, et al. Auditory processing in severely brain injured patients: differences between the minimally conscious state and the persistent vegetative state. *Arch Neurol.* 2004;61:233–238.

263. Kiefer M, Spitzer M. Time course of conscious and unconscious semantic brain activation. *Neuroreport.* 2000;11:2401–2407.

264. Kiefer M. The N400 is modulated by unconsciously perceived masked words: further evidence for an automatic spreading activation account of N400 priming effects. *Brain Res Cogn Brain Res.* 2002;13:27–39.

265. Kiefer M. Repetition-priming modulates category-related effects on event-related potentials: further evidence for multiple cortical semantic systems. *J Cogn Neurosci.* 2005;17:199–211.

266. Jones SJ, Pato MV, Longe O. Auditory information processing in comatose patients: EPs to synthesised 'musical' tones. *Electroencephalogr Clin Neurophysiol Suppl.* 1999;50:402–407.

267. Vaz PM, Jones SJ. Cortical processing of complex tone stimuli: mismatch negativity at the end of a period of rapid pitch modulation. *Brain Res Cogn Brain Res.* 1999;7:295–306.

268. Rappaport M, McCandless KL, Pond W, Krafft MC. Passive P300 response in traumatic brain injury patients. *J Neuropsychiatry Clin Neurosci.* 1991;3:180–185.

269. Connolly JF, Mate-Kole CC, Joyce BM. Global aphasia: an innovative assessment approach. *Arch Phys Med Rehabil.* 1999;80:1309–1315.

270. Kotchoubey B, Lang S, Mezger G, et al. Information processing in severe disorders of consciousness: vegetative state and minimally conscious state. *Clin Neurophysiol.* 2005;116:2441–2453.

271. Kotchoubey B. Event-related potentials, cognition, and behavior: a biological approach. *Neurosci Biobehav Rev.* 2006;30:42–65.

272. Kotchoubey B. Event-related potential measures of consciousness: two equations with three unknowns. *Prog Brain Res.* 2005;150:427–444.

273. Hinterberger T, Wilhelm B, Mellinger J, Kotchoubey B, Birbaumer N. A device for the detection of cognitive brain functions in completely paralyzed or unresponsive patients. *IEEE Trans Biomed Eng.* 2005;52:211–220.

274. Vaitl D, Birbaumer N, Gruzelier J, et al. Psychobiology of altered states of consciousness. *Psychol Bull.* 2005;131:98–127.

275. Neumann N, Kotchoubey B. Assessment of cognitive functions in severely paralysed and severely brain-damaged patients: neuropsychological and electrophysiological methods. *Brain Res Brain Res Protoc.* 2004;14:25–36.

276. Bobes MA, Lopera F, Garcia M, Diaz-Comas L, Galan L, Valdes-Sosa M. Covert matching of unfamiliar faces in a case of prosopagnosia: an ERP study. *Cortex.* 2003;39:41–56.

277. Machado C, Cuspineda E, Valdes P, et al. Assessing acute middle cerebral artery ischemic stroke by quantitative electric tomography. *Clin EEG Neurosci.* 2004;35:116–124.

278. Fernandez-Bouzas A, Harmony T, Fernandez T, et al. Sources of abnormal EEG activity in spontaneous intracerebral hemorrhage. *Clin Electroencephalogr.* 2002;33:70–76.

279. Fernandez-Bouzas A, Harmony T, Fernandez T, et al. Sources of abnormal EEG activity in brain infarctions. *Clin Electroencephalogr.* 2000;31:165–169.

280. Fernandez-Bouzas A, Harmony T, Bosch J, et al. Sources of abnormal EEG activity in the presence of brain lesions. *Clin Electroencephalogr.* 1999;30:46–52.

281. Harmony T, Fernandez T, Silva J, et al. Do specific EEG frequencies indicate different processes during mental calculation? *Neurosci Lett.* 1999;266:25–28.

282. Harmony T, Fernandez-Bouzas A, Marosi E, et al. Frequency source analysis in patients with brain lesions. *Brain Topogr.* 1995;8:109–117.

283. Santiago-Rodriguez E, Harmony T, Graef A, et al. Interictal regional cerebral blood flow and electrical source analysis in patients with complex partial seizures. *Arch Med Res.* 2006;37:145–149.

284. Harmony T, Fernandez T, Gersenowies J, et al. Specific EEG frequencies signal general common cognitive processes as well as specific task processes in man. *Int J Psychophysiol.* 2004;53:207–216.

285. Fernandez-Bouzas A, Harmony T, Fernandez T, Ricardo-Garcell J, Santiago E. Variable resolution electromagnetic tomography (VARETA) in evaluation of compression of cerebral arteries due to deep midline brain lesions. *Arch Med Res.* 2004;35:225–230.

286. Santiago-Rodriguez E, Harmony T, Fernandez-Bouzas A, et al. EEG source localization of interictal epileptiform activity in patients with partial complex epilepsy: comparison between dipole modeling and brain distributed source models. *Clin Electroencephalogr.* 2002;33:42–47.

287. Fernandez-Bouzas A, Harmony T, Fernandez T, Ricardo-Garcell J, Casian G, Sanchez-Conde R. Cerebral blood flow and sources of abnormal EEG activity (VARETA) in neurocysticercosis. *Clin Neurophysiol.* 2001;112:2281–2287.

288. Harmony T, Fernandez T, Fernandez-Bouzas A, et al. EEG changes during word and figure categorization. *Clin Neurophysiol.* 2001;112:1486–1498.

289. Bosch-Bayard J, Valdes-Sosa P, Virues-Alba T, et al. 3D statistical parametric mapping of EEG source spectra by means of variable resolution electromagnetic tomography (VARETA). *Clin Electroencephalogr.* 2001;32:47–61.

290. Jones AK, Kulkarni B, Derbyshire SW. Pain mechanisms and their disorders. *Br Med Bull.* 2003;65:83–93.

291. Jones AK, Kulkarni B, Derbyshire SW. Functional imaging of pain perception. *Curr Rheumatol Rep.* 2002;4:329–333.

292. Kostarczyk E. Recent advances in neonatal pain research. *Folia Morphol (Warsz).* 1999;58:47–56.

293. Howsepian AA. In defense of whole-brain definitions of death. *Linacre Q.* 1998;65:39–61.
294. Kassubek J, Juengling FD, Els T, et al. Activation of a residual cortical network during painful stimulation in long-term postanoxic vegetative state: a ^{15}O-H$_2$O PET study. *J Neurol Sci.* 2003;212:85–91.
295. Katayama Y, Tsubokawa T, Yamamoto T, Hirayama T, Miyazaki S, Koyama S. Characterization and modification of brain activity with deep brain stimulation in patients in a persistent vegetative state: pain-related late positive component of cerebral evoked potential. *Pacing Clin Electrophysiol.* 1991;14:116–121.

8
Brain Death and Organ Transplantation: Ethical Issues

Although there has been a worldwide debate, the idea that a human being with irreversible loss of brain function is dead has been progressively accepted beginning in the early 1960s. The development of the intensive care unit (ICU) facilitated maintaining under mechanical ventilation (and by this means maintaining circulation) patients with complete destruction of their brains.[1–16]

According to Bernat,[17] the new state, termed brain death (BD), shares some attributes seen in alive patients, such as heartbeat, circulation, digestion, and excretion of urine. But it also shares some features of dead patients: unresponsiveness to all stimuli, absence of movement, and apnea. Hence, the application of traditional criteria based on the absence of cardiorespiratory functions is no longer useful in BD patients.[17] Botkin and Post[18] drew an interesting distinction between major and minor clusters of attributes related to life. For example, BD patients retain several attributes associated with life, such as skin color, warm skin, heartbeat, and kidney function. Even subjects who are dead according to the cardiorespiratory standard will preserve vestiges of life attributes for several days: the hair and nails still grow.[9,19]

Moreover, the symbolism that our wishes, our love, and our inspiration are located in our hearts is still fully expressed in different cultures. Poets, writers, and the public usually state expressions such as "I love you from the bottom of my heart." Valentine's Day cards are illustrated with hearts as an expression of love. These examples might explain why there has been reluctance to accept that a patient under ventilatory and cardiocirculatory support and showing a heartbeat, but with an irreversible loss of brain functions, is dead.[1,9,20]

Death as a Biological Notion

Bernat[17,21,22] also states that death is a biological phenomenon, although it has important philosophical, religious, cultural, social, and legal implications. The definition of death that I propose—the irreversible loss of the consciousness, which provides the key human attributes and the highest level of control in the hierarchy of integrating functions within the human organism—has a biological background.[1,9,20] My definition, emphasizing that consciousness provides the main human attributes and is the ultimate integrative function, is more consistent with the biologically based system concepts of Korein et al.[4,23−25] and Bernat et al.[17,21,22,26−32] than the more philosophically based notions of personhood favored by Veatch,[33−41] Barlett and Youngner,[42] Wikler,[43,44] and others.[45−47]

Irreversibility and Potentiality

The issue of irreversibility is directly related to the diagnosis of human death, and it is closely associated with the concept of potentiality, that is, that some patients still have the potential of living. Several authors have emphasized their concerns about accepting that a brain-dead patient is really dead; the same discussion has been extended to non–heart-beating donors.[48−56]

Lizza[57] stated that "potentiality" and "irreversibility" are complementary notions. Hence, if a patient is lacking the potential to retain certain functions, then it is possible to affirm that his or her condition is "irreversible" regarding those functions.

I have no doubt about affirming that the notions of irreversibility and potentiality of being dead or alive are closely related to the historical and technological developments during the era that a human being is living. If we consider a cardiorespiratory definition of death from some decades ago, a cardiac arrhythmia or a cardiorespiratory arrest were almost always irreversible conditions. Nowadays, the development of efficient cardiorespiratory reanimation techniques and the use of mechanical ventilation permit patients in these circumstances to recover, something that was impossible to achieve some years ago.[6]

If we apply in these cases a brain-oriented concept of death, as I have affirmed, a patient does not die in the moment when the heart stops, but rather when anoxia and ischemia are prolonged enough in time to destroy the brain structures.[3−5,7−9,20] Bernat[26] recently affirmed that "the only reliable proof of irreversibility is demonstrating the complete absence of intracranial circulation."

Nonetheless, with the use of a neuroprotective method, a time period of about 5 minutes of asystole, considered enough to produce irreversible destruction of the brain, might be considerably increased without leading to significant nervous system lesions.[58] Hypothermia provides the most effective neuroprotection related to brain resuscitation.[58−66] Safar et al. have stressed that dog experimental models of prolonged exsanguination have resulted in brain and organ preservation during cardiac arrest (no-flow) durations of up to 90 or 120 minutes, at a tympanic temperature of 10 °C.[67,68] This hypothesis is supported by reports of patients undergoing accidental hypothermia (immersion/submersion in cold water, snow avalanche, or prolonged exposure to cold surroundings) combined with circulatory arrest or severe circulatory failure who were rewarmed to normothermia by use of extracorporeal circulation, with good outcome in several cases. In these patients, the neuroprotective effect of accidental hypothermia was initiated very early, even before a complete cardiac arrest had occurred, because of the progressive decrement of body temperature. Hence, in prolonged cardiac arrest with failing conventional measures, rescue by extracorporeal support provides an ultimate therapeutic option with a good outcome in survivors.[58,60,65,69−72]

If, in the future, the brain or the body can be partially of fully reconstituted by some kind of transformed cells[57] or by a not-yet-developed technological device, restoring functions considered now irreversibly lost, physicians would need to change the criteria for death, seeking other functions to assesss irreversibility measure.[6,26,73]

Those who defend cryonic suspension as a method of stabilizing patients who are terminally ill, so that they can be treated in a medical facility of the future, affirm that a disease that would irreversibly cause death at the present time may in the future be cured.[74−80]

Hence, as death will always occur in humans, denying immortality, the criteria for diagnosing death will need to be adapted and correlated to the technological advances of the moment. I still think that at any time in the future brain functions will provide the main human attributes and the highest level of control in the hierarchy of integrating functions within the human organism.[8,9,20,81]

Death: Event vs. Process?

Several authors have discussed whether accepting death is a process or an event. I agree with Bernat,[73] who accepts that death is an event, and that dying and bodily disintegration are processes happening to organisms; dying occurs when the organism is still alive, and disintegration occurs after the organism is already dead.

Brain Death and Organ Transplants

Regarding the "dead donor rule," with the advent of transplant surgery, the definitions and diagnosis of death based on brain formulations have acquired a new urgency.[34,82−88] Nonetheless, as I have already discussed, it is important to point out that the concept of BD as death of the individual did not appear to benefit organ transplantation, but rather was a consequence of the development of intensive care. When French neurophysiologists and neurologists described the death of the nervous system and *coma dépassé*, organ transplant techniques were only in the very early stages of development.[89] As Pallis[90] emphasized, if organ transplant techniques had never been developed, intensive care procedures would have provided the possibility of supplying life support to those patients with destroyed brains and preserved heart function, and physicians would need to face the clinical syndrome of BD.

Nonetheless, educating physicians, nurses, paramedics, and the public about the importance of retrieving organs and tissues from brain-dead cases, for other patients for whom transplantation surgery is the hope for life and good health, is based on the acceptance that BD means the death of the individual.[4−8] This issue has a strong cultural and social background. Japan is an example of a developed country whose religious and social characteristics have presented many obstacles to developing transplantation.[91–95] In contrast, in a developing country, such as Cuba, the education of physicians, nurses, and paramedics and the use of the media to inform public opinion about the concept of death based on neurological grounds and the notion of retrieving organ and tissues for transplant surgery have produced a high rate of family authorization of organ retrievals from a brain-dead donor. In Cuba, although people can express on their identity cards their wishes to be an organ donor, in brain-dead cases physicians always ask permission of patients' relatives for organ retrieval.[3,7,96]

At the Third International Symposium on Coma and Death in Havana in 2000, Alan Shewmon[97,98] presented a striking video of a boy who became brain dead at the age of 4 and who, on ventilator support, showed spontaneous heart-beating 16 years later, although his brain was completely destroyed and liquefied. Thus, it is surely possible to keep "alive" for decades a brain-dead patient, or possibly even a decapitated patient without a functioning heart, with ventilatory assistance and an extracorporeal machine. But are we preserving a corpse or a human being? Thus, I have concluded that a definition of human death should not be related to organ transplants.[6]

Maybe in the near future xenotransplants, clonation, the use of stem cells, or other not-yet-developed methods will allow producing organs and tissues, replacing transplants from brain-dead donors. Nonetheless, physicians will still need to face and deal with the controversial state of brain death.

References

1. Machado C. Cerebral processing in the minimally conscious state. *Neurology.* 2005;65:973–974.

2. Machado C. Havana and the coma and death symposia. *N Engl J Med.* 2004;351:1150–1151.

3. Machado C, Abeledo M, Alvarez C et al. Cuba has passed a law for the determination and certification of death. *Adv Exp Med Biol.* 2004;550:139–142.

4. Korein J, Machado C. Brain death: updating a valid concept for 2004. *Adv Exp Med Biol.* 2004;550:1–14.

5. Machado C, Sherman DL. *Brain Death and Disorders of Consciousness.* New York: Kluwer Academics/Plenum, 2004.

6. Machado C. A definition of human death should not be related to organ transplants. *J Med Ethics.* 2003;29:201–202.

7. Machado C. [Resolution for the determination and certification of death in Cuba]. *Rev Neurol.* 2003;36:763–770.

8. Machado-Curbelo C. [Do we defend a brain oriented view of death?]. *Rev Neurol.* 2002;35:387–396.

9. Machado C. Is the concept of brain death secure? In: Zeman A, Enamorado A, eds. *Ethical Dilemmas in Neurology.* London: WB Saunders, 2000:193–212.

10. Machado-Curbelo C. [A new formulation of death: definition, criteria and diagnostic tests]. *Rev Neurol.* 1998;26:1040–1047.

11. Machado C. Preface. In: Machado C, ed. *Brain Death (Proceedings of the Second International Symposium on Brain Death)*. Amsterdam: Elsevier Science BV, 1995: v–vi.

12. Machado C, García OD, Román JM, Parets J. Four years after the "First International Symposium on Brain Death" in Havana: Could a definitive conceptual re-approach be expected? In: Machado C, ed. *Brain Death (Proceedings of the Second International Symposium on Brain Death)*. Amsterdam: Elsevier Science BV, 1995:1–9.

13. Machado C. A new definition of death based on the basic mechanisms of consciousness generation in human beings. In: Machado C, ed. *Brain Death (Proceedings of the Second International Symposium on Brain Death)*. Amsterdam: Elsevier Science BV, 1995:57–66.

14. Machado C. Una nueva definición de la de muerte según criterios neurológicos. In: Esteban A, Escalante A, eds. *Muerte Encefálica y Donación de Órganos*. Madrid: Comunidad Autónoma de Madrid, 1995:27–51.

15. Machado C. Death on neurological grounds. *J Neurosurg Sci*. 1994;38:209–222.

16. Machado C, García-Tigera J, García O, García-Pumariega J, Román J. Muerte Encefálica. Criterios diagnósticos. Revista Cubana de Medicina 1991;30(3):181–206.

17. Bernat JL. Philosophical and Ethical Aspects of Brain Death. In: Wijdicks EFM, ed. *Brain Death*. Philadelphia: Lippincott Williams & Wilkins, 2001:171–187.

18. Botkin JR, Post SG. Confusion in the determination of death: distinguishing philosophy from physiology. *Perspect Biol Med*. 1992;36:129–138.

19. Pernick MS. Back from the grave: recurring controversies over defining and diagnosing death in history. In: Zaner RM, ed. *Death: Beyond the Whole Brain Criteria*. New York: Kluwer Academics, 1988:17–74.

20. Machado C. Consciousness as a definition of death: its appeal and complexity. *Clin Electroencephalogr*. 1999;30:156–164.

21. Bernat JL. The concept and practice of brain death. *Prog Brain Res*. 2005;150:369–379.

22. Bernat JL. The biophilosophical basis of whole-brain death. *Soc Philos Policy*. 2002;19:324–342.

23. Korein J. The diagnosis of brain death. *Semin Neurol*. 1984;4:52–72.

24. Korein J. Brain death: interrelated medical and social issues. Terminology, definitions, and usage. *Ann N Y Acad Sci*. 1978;315:6–18.

25. Korein J. The problem of brain death: development and history. *Ann N Y Acad Sci*. 1978;315:19–38.

26. Bernat JL. On irreversibility as a prerequisite for brain death determination. *Adv Exp Med Biol*. 2004;550:161–167.

27. Bernat JL. A defense of the whole-brain concept of death. *Hastings Cent Rep*. 1998;28:14–23.

28. Bernat JL. How much of the brain must die in brain death? *J Clin Ethics*. 1992;3:21–26.

29. Bernat JL. Brain death. Occurs only with destruction of the cerebral hemispheres and the brain stem. *Arch Neurol*. 1992;49:569–570.

30. Bernat JL. The definition, criterion, and statute of death. *Semin Neurol*. 1984;4:45–51.

31. Bernat JL, Culver CM, Gert B. Definition of death. *Ann Intern Med*. 1984;100:456.

32. Bernat JL, Culver CM, Gert B. On the definition and criterion of death. *Arch Intern Med*. 1981;94:389–394.

33. Veatch RM. The death of whole-brain death: the plague of the disaggregators, somaticists, and mentalists. *J Med Philos*. 2005;30:353–378.

34. Veatch RM. Abandon the dead donor rule or change the definition of death? *Kennedy Inst Ethics J.* 2004;14:261–276.
35. Veatch RM. The impending collapse of the whole-brain definition of death. *Hastings Cent Rep.* 1993;23:18–24.
36. Veatch RM. Brain death and slippery slopes. *J Clin Ethics.* 1993;4:376.
37. Veatch RM. Defining death at the beginning of life. *Second Opin.* 1989;51–59.
38. Veatch RM. *Death, Dying, and the Biological Revolution. Our Last Responsibility.* New Haven: Yale University Press, 1989.
39. Veatch RM. What it means to be dead. *Hastings Cent Rep.* 1982;12:45–46.
40. Veatch RM. Defining death: the role of brain function. *JAMA.* 1979;242:2001–2002.
41. Veatch RM. The definition of death: ethical philosophical, and policy confusion. *Ann N Y Acad Sci.* 1978;315:307–321.
42. Bartlett ET, Youngner SJ. Human death and the destruction of the neocortex. In: Zaner RM, ed. *Death: Beyond the Whole-Brain Criteria.* New York: Kluwer Academic Publisher, 1988:199–215.
43. Wikler D. Brain death: a durable consensus? *Bioethics.* 1993;7:239–246.
44. Green MB, Wikler D. Brain death and personal identity. *Philos Public Aff.* 1980;9:105–133.
45. Cranford RE, Smith DR. Consciousness: the most critical moral (constitutional) standard for human personhood. *Am J Law Med.* 1987;13:233–248.
46. Truog RD, Cochrane TI. The truth about "donation after cardiac death". *J Clin Ethics.* 2006;17:133–136.
47. Truog RD, Fackler JC. Rethinking brain death. *Crit Care Med.* 1992;20:1705–1713.
48. Arnold RM, Youngner SJ. Time is of the essence: the pressing need for comprehensive non-heart-beating cadaveric donation policies. *Transplant Proc.* 1995;27:2913–2917.
49. Youngner SJ, Arnold RM. Ethical, psychosocial, and public policy implications of procuring organs from non-heart-beating cadaver donors. *JAMA.* 1993;269:2769–2774.
50. Arnold RM, Youngner SJ. Back to the future: obtaining organs from non-heart-beating cadavers. *Kennedy Inst Ethics J.* 1993;3:103–111.
51. Revelly JP, Imperatori L, Maravic P, Schaller MD, Chiolero R. Are terminally ill patients dying in the ICU suitable for non-heart beating organ donation? *Intensive Care Med.* 2006.
52. Hughes RD, Mitry RR, Dhawan A et al. Isolation of hepatocytes from livers from non-heart-beating donors for cell transplantation. *Liver Transpl.* 2006;12(5):713–717.
53. Hirota M, Ishino K, Fukumasu I et al. Prediction of functional recovery of 60–minute warm ischemic hearts from asphyxiated canine non-heart-beating donors. *J Heart Lung Transplant.* 2006;25:339–344.
54. Deveaux TE. Non-heart-beating organ donation: Issues and ethics for the critical care nurse. *J Vasc Nurs.* 2006;24:17–21.
55. White SA, Prasad KR. Liver transplantation from non-heart beating donors. *BMJ.* 2006;332:376–377.
56. Nunez JR, Del RF, Lopez E, Moreno MA, Soria A, Parra D. Non-heart-beating donors: an excellent choice to increase the donor pool. *Transplant Proc.* 2005;37:3651–3654.
57. Lizza JP. Potentiality, irreversibility, and death. *J Med Philos.* 2005;30:45–64.
58. Machado C. Randomized clinical trial of magnesium, diazepam, or both after out-of-hospital cardiac arrest. *Neurology.* 2003;60:1868–1869.

59. Brieva J, McFadyen B, Rowley M. Severe hypothermia: challenging normal physiology. *Anaesth Intensive Care.* 2005;33:662–664.
60. Schewe JC, Heister U, Fischer M, Hoeft A. [Accidental urban hypothermia. Severe hypothermia of 20.7° C]. *Anaesthesist.* 2005;54:1005–1011.
61. Wijdicks EF. Induced hypothermia in neurocatastrophes: feeling the chill. *Rev Neurol Dis.* 2004;1:10–15.
62. Safar P. Mild hypothermia in resuscitation: a historical perspective. *Ann Emerg Med.* 2003;41:887–888.
63. Deseano-Estudillo JL, Castanon-Gonzalez JA, Carbajal-Ramirez A, Castrejon-Roman H, Leon-Gutierrez MA. [Cerebral blood flow velocity spectrum by transcranial doppler ultrasound in patients with brain death clinical criteria.]. *Gac Med Mex.* 2003;139:535–538.
64. Zhou L, Shao Z, Ou S. Cryoanalgesia: electrophysiology at different temperatures. *Cryobiology.* 2003;46:26–32.
65. Brat R, Suk M, Barta J et al. [Resuscitation of a patient with deep hypothermia using extracorporeal circulation]. *Rozhl Chir.* 2002;81:279–281.
66. Danzl D. Hypothermia. *Semin Respir Crit Care Med.* 2002;23:57–68.
67. Safar PJ, Tisherman SA. Suspended animation for delayed resuscitation. *Curr Opin Anaesthesiol.* 2002;15:203–210.
68. Safar P, Behringer W, Bottiger BW, Sterz F. Cerebral resuscitation potentials for cardiac arrest. *Crit Care Med.* 2002;30:S140–S144.
69. Freye E. Cerebral monitoring in the operating room and the intensive care unit—an introductory for the clinician and a guide for the novice wanting to open a window to the brain. Part II: sensory-evoked potentials (SSEP, AEP, VEP). *J Clin Monit Comput.* 2005;19:77–168.
70. Farstad M, Andersen KS, Koller ME, Grong K, Segadal L, Husby P. Rewarming from accidental hypothermia by extracorporeal circulation. A retrospective study. *Eur J Cardiothorac Surg.* 2001;20:58–64.
71. Farbrot K, Husby P, Andersen KS, Grong K, Farstad M, Koller ME. [Rewarming of patients with accidental hypothermia with the help of heart-lung machine]. *Tidsskr Nor Laegeforen.* 2000;120:1854–1857.
72. Cortes J, Galvan C, Sierra J, Franco A, Carceller J, Cid M. [Severe accidental hypothermia: rewarming by total cardiopulmonary bypass]. *Rev Esp Anestesiol Reanim.* 1994;41:109–112.
73. Bernat JL. Philosophical and Ethical Aspects of Brain Death. In: Wijdicks EFM, ed. *Brain Death.* Philadelphia: Lippincott Williams & Wilkins, 2001:171–88.
74. Shermer M. Nano nonsense and cryonics. *Sci Am.* 2001;285:29.
75. Pommer RW, III. Donaldson v. Van de Kamp: cryonics, assisted suicide, and the challenges of medical science. *J Contemp Health Law Policy.* 1993;9:589–603.
76. Merkle RC. The technical feasibility of cryonics. *Med Hypotheses.* 1992;39:6–16.
77. LaBouff JP. "He wants to do what?" Cryonics: issues in questionable medicine and self-determination. *Santa Clara Comput High Technol Law J.* 1992;8:469–498.
78. Smith GP. The iceperson cometh: cryonics, law and medicine. *Health Matrix.* 1983;1:23–35.
79. Kastenbaum R, Ettinger R. Cryonic suspension: an Omega interview with R.C.W. Ettinger. *Omega (Westport).* 1994;30:159–171.
80. Ettinger R. A matter of life and death. *Analog Sci Fict Sci Fact.* 1979;99:67–70.
81. Machado C, García OD, Gutiérrez J, Portela L, García MC. Heart rate variability in comatose and brain-dead patients. *Clin Neurophysiol.* 2005;116:2859–2860.

82. Arbour R. Clinical management of the organ donor. *AACN Clin Issues.* 2005;16:551–580.

83. Siminoff LA, Burant C, Youngner SJ. Death and organ procurement: public beliefs and attitudes. *Kennedy Inst Ethics J.* 2004;14:217–234.

84. Crowley-Matoka M, Arnold RM. The dead donor rule: how much does the public care . . . and how much should we care? *Kennedy Inst Ethics J.* 2004;14:319–332.

85. Campbell CS. Harvesting the living?: separating "brain death" and organ transplantation. *Kennedy Inst Ethics J.* 2004;14:301–318.

86. Shewmon DA. The dead donor rule: lessons from linguistics? *Kennedy Inst Ethics J.* 2004;14:277–300.

87. Veatch RM. The dead donor rule: true by definition. *Am J Bioeth.* 2003;3:10–11.

88. Komokata T, Nishida S, Ganz S, Suzuki T, Olson L, Tzakis AG. The impact of donor chemical overdose on the outcome of liver transplantation. *Transplantation.* 2003;76:705–708.

89. Machado C. The first organ transplant from a brain-dead donor. *Neurology.* 2005;64:1938–1942.

90. Pallis C. Brainstem death. In: Braakman R, ed. *Handbook of Clinical Neurology: Head Injury.* Amsterdam: Elsevier Science BV, 1990:441–496.

91. Feldman EA. Brain death: the Japanese controversy. In: Machado C, ed. *Brain Death (Proceedings of the Second International Symposium on Brain Death).* Amsterdam: Elsevier Science BV, 1995:265–286.

92. Teraoka S. [Organ sharing network for organ transplantation from heart-beating deceased donors]. *Nippon Rinsho.* 2005;63:1879–1887.

93. Ota K. Current status of organ transplants in Asian countries. *Transplant Proc.* 2004;36:2535–2538.

94. Handa H. [Controversy over brain death in Japan]. *J UOEH.* 1990;12:309–314.

95. Shiogai T, Takeuchi K. [Brain death in a neurosurgical unit in Japan]. *No To Shinkei.* 1984;36:781–787.

96. Machado C. Determination of death. *Acta Anaesthesiol Scand.* 2005;49:592–593.

97. Shewmon DA. A critical analysis of conceptual domains of the vegetative state: sorting fact from fancy. *NeuroRehabilitation.* 2004;19:343–347.

98. Shewmon DA, Holmes GL, Byrne PA. Consciousness in congenitally decorticate children: developmental vegetative state as self-fulfilling prophecy. *Dev Med Child Neurol.* 1999;41:364–374.

9
Legal Considerations on the Determination and Certification of Human Death

During the last several decades physicians and the public have needed urgent changes in the legal codes for accepting brain death (BD) as death, in order to obtain organs from heart-beating donors. As transplantation has become increasingly more effective, the demand for organs has outpaced the supply. Patients in need of organ transplants are dying needlessly because of the paucity of organs. The needs of tissues and organs for treatment through transplantation and the demand for organs quickly outpaced the supply; consequently, it became evident that standards and priorities needed to be established in order to allocate the increasingly scarce number of organs for transplantation. The transplant community continues to grapple with the organ-sharing shortage, to find new and inventive ways to increase the donor pool, and to allocate organs in a fair and equitable way based on sound scientific decision making, while enhancing the science of transplantation. The "dead donor rule" requires that donors must first be declared dead.[1-9]

Legal criteria vary from place to place. Most organ donation for organ transplantation is done in the setting of BD. In some countries (for example, Belgium, Poland, and Portugal) everyone is automatically an organ donor unless you specifically request not to be one.[10-27]

A landmark approach to this area in the United States occurred in 1980, when the Uniform Law Commissioners adopted the Uniform Determination of Death Act (UDODA).[28] According to the UDODA, "An individual who has sustained either:

1. irreversible cessation of circulatory and respiratory functions, or
2. irreversible cessation of all functions of the entire brain, including the brainstem, is dead."

The UDODA was subsequently endorsed by the American Medical Association, the American Bar Association, the President's Commission for the Study of Ethical Problems in Medicine and Biomedical and Behavioral Research, the American Academy of Neurology, and the American Electroencephalographic Society. Recommendations of the Task Force, therefore, included the development of uniform standards for organ procurement, and implementation of the UDODA in all states.[28-32]

In 1984, the United States Congress passed one of the most significant legislative initiatives ever introduced into the field of transplantation, the National Organ Transplant Act (NOTA), which established a 25-member multidisciplinary Task Force on Organ Transplantation, charged with conducting a comprehensive examination of the medical, legal, ethical, social, and economic issues arising from the practices of human organ donation, recovery, and transplantation.[33]

When BD is declared, the transplantable organs can remain viable for an extended period via the use of a ventilator, continuing the process of providing nutrients. After the declaration of death, the family can be given time to accept the occurrence of death before being approached for organ donation, a process called decoupling. It also provides time for families to consider the donation option.[34,35]

Hence, in several countries and states legislation was needed to establish BD as a legal criterion for organ donation. For this reason, most codes legalizing BD are usually sections of transplant laws.[13,35-50] Thus, a conceptual and practical controversy emerged: if BD cases were not useful as organ donors, they were usually kept on life support until cardiac arrest occurred.[13,46,51,52]

The Cuban National Commission for the Determination and Certification of Death has accepted a neurological view of death and stated that any legal code of death should be completely separated from any norm governing organ transplants. Moreover, as was discussed in Chapter 3, although the commission accepted a definition of death based only on the irreversible loss of brain functions, there are several ways of diagnosing death, according to the scenario where death occurs.[13,46,51-55]

Cuban Law for the Determination of Death

The Civil Code

The Cuban Civil Code has undergone several historical changes regarding the determination of death.[56] The Spanish Civil Code, established in Cuba since 1889, stated in Article 31: "Death occurs when the person is extinguished." The new Civil Code (Cuban Law 1175, March 9, 1965) stated: "Death is the loss of every sign of life after a living birth."

The development of Article 26.1 of the present Civil Code (July 1987) stated: "Physicians are the only professionals authorized to diagnose and certify death according to a norm established by the Ministry of Public Health."

Distinguishing Features of Article 26.1 of the Current Cuban Civil Code[13,56,57]

- It does not define death.
- It specifies that physicians are the only professionals authorized to diagnose and certify death.

ISSN 0864-0793

GACETA OFICIAL
DE LA REPUBLICA DE CUBA

MINISTERIO DE JUSTICIA

EDICION ORDINARIA LA HABANA, VIERNES 21 DE SEPTIEMBRE DEL 2001   AÑO XCIX

SALUD PÚBLICA'
RESOLUCIÓN MINISTERIAL N° 90

FIGURE 9.1. Cuban law for the determination and certification of death.

- It specifies that the Ministry of Public Health must provide a norm for the determination of death (clinical and instrumental criteria).
- The Cuban Parliament gave full responsibility to the Ministry of Public Health to legalize the determination and certification of death.

Law Within a Ministry

According to the Cuban Parliament, a resolution is a law that legalizes a working standard within any ministry and that could be signed and changed by the minister in function. Therefore, the Ministry of Public Health needed to respond to the present Civil Code, Article 26.1, by writing a resolution on this subject.[56,57]

For this reason, the Ministry of Public Health organized the National Commission for the Determination and Certification of Death at the beginning of the 1990s. The commission was composed of representatives from multiple disciplines, including various medical and legal specialties. After several versions, the commission wrote the final resolution, signed by the Minister of Public Health on August 27, 2001 (Fig. 9.1).[57]

The main body of the norm contained the following points[13,57]:

- Physicians diagnose death by documenting the "signs of death."
- When the diagnosis of death is based on the irreversible loss of whole brain functions, medical specialists need to be accredited by the National Commission for the Determination and Certification of Death.
- Death is certified by the physician who diagnoses it. Physicians will register the moment of death upon completing the diagnostic procedure.

- The commission will annually review the diagnostic criteria of death and will propose changes or amendments, according to medical and technological advances in this area.

Diagnosis of Death

The National Commission accepted only one kind of death, based on the *irreversible loss of brain functions*. As the Civil Code did not require a definition, the commission followed this norm and presented not a concept but rather the ways of diagnosing death. However, discussions left very clear that although there is only one kind of death, there are several ways of diagnosing it, according to the place and environment where death occurs.[13,51,52,56–58]

Nonetheless, the commission adopted the view that the irreversible loss of cardiocirculatory and respiratory functions can only cause death when ischemia and anoxia are prolonged enough to produce an irreversible destruction of the brain.[13,51–59]

Physicians diagnose death by finding the "signs of death." Signs I and II correspond to the classic respiratory and cardiocirculatory functions. Signs III to VIII are related to forensic circumstances. Sign IX corresponds to the BD diagnosis (see Chapter 4).

Conclusion

Any legal code of death should be completely separate from any norm governing organ transplants. Moreover, although there is only one kind of human death, based on the irreversible loss of brain functions, there are several ways of diagnosing death, according to the scenario where death occurs. Nonetheless, it does not mean that there are different kinds of deaths.

References

1. Shewmon DA. The dead donor rule: lessons from linguistics? *Kennedy Inst Ethics J.* 2004;14:277–300.
2. Veatch RM. Abandon the dead donor rule or change the definition of death? *Kennedy Inst Ethics J.* 2004;14:261–276.
3. Crowley-Matoka M, Arnold RM. The dead donor rule: how much does the public care... and how much should we care? *Kennedy Inst Ethics J.* 2004;14:319–332.
4. Fost N. Reconsidering the dead donor rule: is it important that organ donors be dead? *Kennedy Inst Ethics J.* 2004;14:249–260.
5. Veatch RM. The dead donor rule: true by definition. *Am J Bioeth.* 2003;3:10–11.
6. Siminoff LA. The dead donor rule: not dead yet. *Am J Bioeth.* 2003;3:30.
7. Shelton W. Respect for donor autonomy and the dead donor rule. *Am J Bioeth.* 2003;3:20–21.
8. Pasquerella L, Smith S, Ladd R. Infants, the dead donor rule, and anencephalic organ donation: should the rules be changed? *Med Law.* 2001;20:417–423.

9. Robertson JA. The dead donor rule. *Hastings Cent Rep.* 1999;29:6–14.

10. Bernat JL. Ethical and legal aspects of the emergency management of brain death and organ retrieval. *Emerg Med Clin North Am.* 1987;5:661–676.

11. Bernat JL, D'Alessandro AM, Port FK et al. Report of a national conference on donation after cardiac death. *Am J Transplant.* 2006;6:281–291.

12. Bernat JL. Philosophical and Ethical Aspects of Brain Death. In: Wijdicks EFM, ed. *Brain Death.* Philadelphia: Lippincott Williams & Wilkins, 2001:171–88.

13. Machado C. [Resolution for the determination and certification of death in Cuba]. *Rev Neurol.* 2003;36:763–770.

14. van Deynse D, Dumont V, Lecomte C et al. Non-heart-beating donor, renewal of an old procedure: a Belgian experience. *Transplant Proc.* 2005;37:2821–2822.

15. Randhawa G. Procuring organs for transplantation—a European perspective. *Eur J Public Health.* 1998;8:299–304.

16. Spital A. Consent for organ donation: today and tomorrow. *Semin Dial.* 1993;6:264–267.

17. Nys H. Therapeutic use of human organs and tissues under Belgian law. *Med Law.* 1993;12:131–135.

18. Toussaint C. The new act regulating organ retrieval for transplantation in Belgium. *Contrib Nephrol.* 1989;71:141–144.

19. Nojszewska M. [Determination of brain death]. *Neurol Neurochir Pol.* 2002;36:91–104.

20. Andres J, Macheta A. [Persistent vegetative state: medical, moral, legal and economic aspects]. *Folia Med Cracov.* 1998;39:73–77.

21. Lao M, Rowinski W, Walaszewski J, Szmidt J. Renal transplantation in Warsaw: a registry report. *Transplant Proc.* 1996;28:3497–3503.

22. Michalak G, Danielewicz R, Ostrowski K, Barcikowska B, Walaszewski JE, Rowinski WA. Is the decreased number of renal transplants in Warsaw caused by a shortage of potential organ donors?: an analysis of hospital deaths in Warsaw in 1993. *Transplant Proc.* 1996;28:270–271.

23. Rowinski W, Ostrowski K, Adadynski L et al. Factors limiting renal transplantation program in Poland. *Ann Transplant.* 1996;1:18–22.

24. Baran E, Sych M. [Medico-legal circumstances of organ transplantation in Poland (history and present state)]. *Przegl Lek.* 1996;53:883–885.

25. Rowinski WA, Walaszewski J, Madej K, Szmidt J, Lao M. Demand and supply of kidneys for transplantation in Poland. *Transplant Proc.* 1993;25:3129.

26. Sych M. [Committee of Medical Sciences of the Polish Academy of Sciences in Cracow. Report on a closed session entitled: "Criteria of sudden intravital brain death" and held on December 17, 1974]. *Folia Med Cracov.* 1976;18:311–314.

27. Pena JR. [Kidney transplant: the current situation in Portugal, 1989]. *Acta Med Port.* 1989;suppl 2:S11–S15.

28. Uniform determination of Death Act. *Conn Med.* 1982;46:259–260.

29. Recommendations for use of uniform nomenclature pertinent to patients with severe alterations in consciousness. American Congress of Rehabilitation Medicine. *Arch Phys Med Rehabil.* 1995;76:205–209.

30. Day L. Practical limits to the Uniform Determination of Death Act. *J Neurosurg Nurs.* 1995;27:319–322.

31. Determination of death (Uniform Determination of Death Act of 1981); natural death (Natural Death Act of 1981). *Lexis DC Code DC.* 1982;Sect. 6.2401 6.2421 to 6.2430 amended Feb 1982:Unknown.

32. Abram MB. The need for uniform law on the determination of death. *NY Law Sch Law Rev.* 1982;27:1187–1205.

33. National Organ Transplant Act: Public Law 98–507. *US Statut Large.* 1984;98:2339–2348.

34. Adams RK. Live organ donors and informed consent. A difficult minuet. *J Leg Med.* 1987;8:555–586.

35. McDonald JC. The National Organ Procurement and Transplantation Network. *JAMA.* 1988;259:725–726.

36. Arpesella G, Gherardi S, Bombardini T, Picano E. Recruitment of aged donor heart with pharmacological stress echo. A case report. *Cardiovasc Ultrasound.* 2006;4:3.

37. Asano Y, Ashikari J. [Report from the Japan organ transplant network]. *Nippon Jinzo Gakkai Shi.* 2005;47:517–523.

38. Perez-Tamayo R. [The law, the medical ethic, and the transplants]. *Rev Invest Clin.* 2005;57:170–176.

39. Hattori M, Iwaki Y. [Organs shortage in the United States]. *Nippon Rinsho.* 2005;63:1888–1897.

40. Muller C, Vernet D, Moretti D, Klinger S, Vernet JP. [Organ donation in Switzerland]. *Ther Umsch.* 2005;62:437–442.

41. Monnier M, Weber T. [The Transplantation Law of October 8, 2004]. *Ther Umsch.* 2005;62:433–435.

42. Dimond B. Law concerning organ transplants and dead donors in the UK. *Br J Nurs.* 2005;14:47–48.

43. Murray JE. The first successful organ transplants in man. *J Am Coll Surg.* 2005;200:5–9.

44. Slooff MJ, Kazemier G. [Liver transplantation from a living donor]. *Ned Tijdschr Geneeskd.* 2004;148:2257–2259.

45. Peron AL, Rodrigues AB, Leite DA et al. Organ donation and transplantation in Brazil: university students' awareness and opinions. *Transplant Proc.* 2004;36:811–813.

46. Machado C. A definition of human death should not be related to organ transplants. *J Med Ethics.* 2003;29:201–202.

47. Nemec P. [Ethical questions in organ transplantation]. *Vnitr Lek.* 2002;48:667–670.

48. Bechstein WO, Wullstein C. [Transplantation and outcome quality]. *Chirurg.* 2002;73:576–581.

49. Luna-Zaragoza D, Reyes-Frias ML. Donation transplants and tissue banking in Mexico. *Cell Tissue Bank.* 2001;2:255–259.

50. Jennett B. Transplants—are the donors really dead? *Br Med J.* 1980;281:1423–1424.

51. Machado C. Determination of death. *Acta Anaesthesiol Scand.* 2005;49:592–593.

52. Machado C, Abeledo M, Alvarez C et al. Cuba has passed a law for the determination and certification of death. *Adv Exp Med Biol.* 2004;550:139–142.

53. Machado C, Garcia OD, Gutierrez J, Portela L, Garcia MC. Heart rate variability in comatose and brain-dead patients. *Clin Neurophysiol.* 2005;116:2859–2860.

54. Machado-Curbelo C. [Do we defend a brain oriented view of death?]. *Rev Neurol.* 2002;35:387–396.

55. Machado C. Consciousness as a definition of death: its appeal and complexity. *Clin Electroencephalogr.* 1999;30:156–164.

56. Ley No. 59 del 16 de Julio de 1987 (Codigo Civil). Edicion Extraordinaria no. 9. *Gaceta Oficial de la Republica de Cuba.* Ciudad de La Habana: Ministerio de Justicia, 1987:39–81.

57. Resolucion no. 90 de Salud Publica. Edicion Ordinaria del 21 de Septiembre del 2001. Ciudad de La Habana: Ministerio de Justicia, 2001.

58. Machado C, García OD. Guidelines for the determination of brain death. In: Machado C, ed. *Brain Death (Proceedings of the Second International Symposium on Brain Death)*. Amsterdam: Elsevier Science BV, 1995:75–80.
59. Machado C. Is the concept of brain death secure? In: Zeman A, Enamorado A, eds. *Ethical Dilemmas in Neurology*. London: WB Saunders, 2000:193–212.

Index

Page numbers followed by f indicate figures; t, tables.

Printed in the United States of America